Cooperative Management

Series Editors

The Book Series on Cooperative Management provides an invaluable forum for creative and scholarship work on cooperative economics, organizational, financial and marketing aspects of business cooperatives and development of cooperative communities throughout the Mediterranean region and worldwide. The main objectives of this book series are to advance knowledge related to cooperative entrepreneurship as well as to generate theoretical knowledge aiming to promoting research within various sectors wherein cooperatives operate (agriculture, banking, real estate, insurance, and other forms). Scholarly edited volumes and monographs should relate to one of these areas, should have a theoretical and/or empirical problem orientation, and should demonstrate innovation in theoretical and empirical analyses, methodologies, and applications. Analyses of cooperative economic problems and phenomena pertinent to managerial research, extension, and teaching (e.g., case studies) regarding cooperative entrepreneurship are equally encouraged.

Konstadinos Mattas • George Bacurakis
Constantin Zopounidis • Christos Staboulis
Editors

Biodiversity in Agri-Food Systems

Food Security and Climate Change Mitigation

Springer

Editors
Konstadinos Mattas
Department of Agricultural Economics
Aristotle University of Thessaloniki
Thessaloniki, Greece

Constantin Zopounidis
School of Production Engineering and Management
Technical University of Crete
Chania, Greece

George Baourakis
Business Economics and Management Department
Mediterranean Agronomic Institute of Chania
Chania, Greece

Christos Staboulis
Department of Agricultural Economics
Aristotle University of Thessaloniki
Thessaloniki, Greece

ISSN 2364-401X ISSN 2364-4028 (electronic)
Cooperative Management
ISBN 978-3-032-06384-7 ISBN 978-3-032-06385-4 (eBook)
https://doi.org/10.1007/978-3-032-06385-4

This Springer imprint is published by the registered company Springer Nature Switzerland AG
The registered company address is: Gewerbestrasse 11, 6330 Cham, Switzerland

Preface

The global agri-food system is at a crossroads. Climate change, shifting consumer expectations, sustainability imperatives, and economic uncertainties are reshaping how food is produced, distributed, and consumed. The interconnected challenges of food security, supply chain resilience, sustainable business models, and environmental sustainability require innovative strategies to ensure the long-term viability of agri-food chains. This collective volume brings together diverse perspectives and empirical studies that contribute to understanding and addressing these pressing issues.

The first chapter explores the critical role of sustainable consumption in shaping supply chains within the European Union. Through a large-scale survey of consumer behavior, the study highlights how purchasing choices and willingness to pay for sustainable products influence supply chain management. This chapter underscores the pivotal role of consumer awareness and government policy in fostering sustainable supply chain practices, emphasizing the necessity for collaboration between policymakers, businesses, and consumers.

Supply chain resilience is a central concern in the second chapter, which examines the Scottish pig industry's response to economic and geopolitical shocks. By analyzing the impact of events such as Brexit, the COVID-19 pandemic, and supply chain disruptions, the study highlights the importance of horizontal collaboration, particularly through producer cooperatives. The chapter contributes to a broader understanding of how collective action can mitigate risks and enhance long-term sustainability in the agri-food sector.

The role of alternative food systems is further explored in the third chapter, which delves into the expanding field of wine tourism. The study identifies key motivations driving tourists' choices and their impact on local food economies. By analyzing consumer behavior, destination satisfaction, and loyalty, this chapter presents valuable insights into how experiential tourism can foster sustainable economic models in rural regions. The discussion highlights the broader theme of integrating tourism, local food production, and cultural identity within sustainable agri-food systems.

Water resource management is a crucial component of agricultural sustainability; the fourth chapter presents an evaluation of water balance across selected watersheds in Africa. By utilizing the Thornthwaite-Mather method, the study assesses water availability for irrigation and agricultural planning. Findings underscore the disparities in water resources between East and West Africa, emphasizing the need for region-specific water management strategies to optimize agricultural productivity and enhance climate resilience.

Sustainability in business operations takes center stage in the fifth chapter, which applies the Triple Bottom Line (TBL) framework to the fresh produce industry in Spain. The chapter demonstrates how companies can transition from traditional economic models to sustainability-driven business practices. It highlights how integrating socio-environmental standards into business models enhances supply chain transparency, resilience, and competitive advantage, reinforcing the book's broader discussion on sustainable value creation.

The final chapter examines the quality and competitiveness of small-scale dairy farms in Hungary. It explores consumer trust in food safety, challenges faced by farmers, and the role of HACCP processes in ensuring quality control. This study adds a crucial dimension to the book by addressing the intersection of food safety, consumer perceptions, and supply chain integration for small producers.

Taken together, these contributions provide a comprehensive examination of how agri-food chains can be transformed to meet the dual challenges of climate change and food insecurity. The insights presented in this volume offer valuable guidance for policymakers, industry stakeholders, and researchers seeking to build more resilient, sustainable, and equitable food systems. As climate change and global uncertainties continue to reshape agri-food chains, the findings in this book underscore the importance of innovation, collaboration, and evidence-based policymaking in ensuring a sustainable food future.

We would like to thank Maria Verivaki, English professor, for proofreading. We extend our appreciation to the authors and referrers of these chapters, and Springer Academic Publications for their assistance in producing this book.

Thessaloniki, Greece — Konstadinos Mattas
Chania, Greece — George Baourakis
Chania, Greece — Constantin Zopounidis
Thessaloniki, Greece — Christos Staboulis

Contents

Chapter 1
The Impact of Sustainable Consumption on Supply Chain Management: An EU-Wide Study

Foivos Anastasiadis, Naoum Tsolakis, Nikolaos Syndoukas, Angelos Stamos, Anastasios Kyriakidis, Anastasia Mousiadou, Nina Louvrou, Ioannis Manikas, and Konstadinos Mattas

1.1 Introduction

Sustainable consumption is a crucial focus in the effort to preserve the environment and promote social equity. As the global population grows and resource demands increase, there is an urgent need to change consumption patterns. The United Nations has highlighted sustainable consumption as a vital part of its Sustainable

F. Anastasiadis (✉)
Department of Agribusiness and Supply Chain Management, Agricultural University of Athens, Athens, Greece
e-mail: foivos@aua.gr

N. Tsolakis
Department of Supply Chain Management, International Hellenic University, Thessaloniki, Greece
e-mail: ntsolakis@ihu.gr

N. Syndoukas · K. Mattas
Department of Agricultural Economics, Aristotle University of Thessaloniki, Thessaloniki, Greece
e-mail: nikosynd@agro.auth.gr; mattas@auth.gr

A. Stamos
Department of Management, Society and Communication, Copenhagen Business School, Copenhagen, Denmark
e-mail: asta.msc@cbs.dk

A. Kyriakidis · A. Mousiadou · N. Louvrou
Research Department, White Research SRL, Thessaloniki, Belgium
e-mail: akyriakidis@white-research.eu; amousiadou@white-research.eu; nlouvrou@white-research.eu

I. Manikas
Faculty of Agrobiology, Food, and Natural Resources, Czech University of Life Sciences Prague, Prague, Czech Republic
e-mail: manikas@af.czu.cz

K. Mattas et al. (eds.), *Biodiversity in Agri-Food Systems*, Cooperative Management, https://doi.org/10.1007/978-3-032-06385-4_1

Development Goals, particularly Goal #12, which aims to ensure sustainable consumption and production patterns (UN DESA, 2023). This Goal is important for the well-being of current and future generations, as unsustainable consumption practices can deplete natural resources and worsen climate change. The environmental impacts of overconsumption are undeniable, contributing to significant issues such as climate change, biodiversity loss, and pollution. A report by the United Nations Environment Programme stressed that current consumption levels exceed the planet's capacity to regenerate resources, leading to a cycle of waste and environmental degradation (Wang et al., 2019). For instance, food waste alone accounts for approximately 931 million tons annually, illustrating the inefficiencies and unsustainable practices in current consumption patterns (UN DESA, 2023). These alarming statistics underscore consumers' need to adopt more responsible consumption behaviours prioritizing sustainability.

The significance of sustainable consumption is widespread; nonetheless, there remains room for improvement in its integration within the operations management domain. For example, emerging economies are falling behind in adopting Sustainable Supply Chain Management (SSCM) approaches (Sánchez-Flores et al., 2020). Many businesses follow traditional SC (supply chain) models, prioritising efficiency and cost-effectiveness over sustainability. The latter *modus operandi* fuels major sustainability concerns as it impedes efforts to promote responsible consumption and address the broader environmental challenges. Research indicates that integrating sustainable consumption principles into SC strategies can drive significant systemic change. Companies should embrace sustainable consumption practices to reduce their ecological footprints and enhance their competitive advantage (De Oliveira et al., 2022). However, the existing body of literature often treats sustainable consumption and SCM as separate elements and overlooks the potential synergies that could arise from their integration.

It is essential to address this issue to advance both sustainable consumption and SCM practices. An end-to-end SC perspective provides a holistic view of sustainability challenges, allowing stakeholders to identify opportunities for improvement and collaboration. Furthermore, consumers increasingly demand transparency and accountability from businesses regarding their sustainability practices (Anastasiadis et al., 2024), forcing companies to adjust their SCs accordingly (Anastasiadis et al., 2022).

Therefore, this paper aims to explore the interplay between sustainable consumption and SCM within the context of the European Union. By examining how sustainable consumption patterns can inform and transform SC strategies, this research seeks to contribute to the broader discourse on sustainable development and provide actionable insights for policymakers and businesses alike. Addressing the integration of sustainable consumption into SCM not only aligns with global sustainability goals but also fosters a more resilient and equitable economy.

1.2 Literature Review

1.2.1 Supply Chain Management and Sustainability

The necessity of creating precise and efficient supply and distribution channels has grown over the past few decades. As a result, the idea of SCM has gained prominence in the related literature. The food SC is defined as "*a set of interdependent companies that work closely together to manage the flow of goods and services along the value-added chain of agricultural and food products, in order to realise superior customer value at the lowest possible costs*" (Folkerts & Koehorst, 1997). Conversely, SCM integrates the above-mentioned activities and end-user business processes to improve the network relationships, aiming to provide products, services, and information that add value to customers (Cooper et al., 1997). In the early stages of its development, societal and environmental concerns were not among the primary issues addressed by SCM. Instead, economic factors were the driving force behind the focus on SCM, with the primary objective being the reduction of financial losses through the minimisation of waste and the reduction of costs in other related processes (Sarkis et al., 2011).

Over the last few years, there has been a growing need to integrate the environmental and social aspects into SCM. As a result, stakeholders within SCs have increasingly adopted the concept of sustainability in their operations. This has led to the evolution of the term SCM (supply chain management) into SSCM (sustainable supply chain management), expanding the sustainable development issues from the organisation to the SC level (Carter & Rogers, 2008). The most commonly accepted definition of sustainable development describes it as "*the development that meets the needs of the present without compromising the ability of future generations to meet their own needs, with a focus on environmental, social, and economic dimensions*" (Mitlin, 1992). Sustainability is firmly grounded in both the natural and social sciences. Advancements across various natural science disciplines enable us to grasp the impacts and interactions of diverse human activities on the natural environment, affecting the quality of life today and in the future. Equally vital are the social sciences, which shape our understanding and approach to sustainability through exploring cultural norms, individual and collective behaviours, as well as governmental and community roles. According to Linton, Klassen and Jayaraman (2007), anthropology, psychology, and sociology collaborate with the natural sciences, informing policy development and management strategies.

When all the aforementioned elements are combined, the concept of SSCM is created. This involves managing the flows of capital, information, and materials (Mentzer et al., 2001) while considering the environmental, social, and economic aspects of sustainable development as stakeholders and customers require. In SSCs, environmental and social criteria must be met while maintaining competitiveness through customer satisfaction and adherence to relevant economic criteria (Seuring & Müller, 2008). It is, therefore, evident that efforts are no longer focused solely on improving processes in the upstream echelons of a SC (farms, processors, and retailers), but the satisfaction and involvement of end recipients/consumers downstream the SC are now considered equally important. Studies such as those by Poore and Nemecek (2018) and Zhang, Gu and Jahromi (2019) emphasised the role of consumers in adopting effective SSCM to mitigate and reduce the negative environmental impacts through their purchasing choices, behaviours, and habits (Hu et al., 2019).

1.2.2 Sustainable Consumption

Sustainable food consumption is a notion that encompasses three fundamental dimensions: economic, ecological, and social (Lim & Londob, 2018). The economic dimension concentrates on ensuring food affordability, equitable access, and fair compensation for all parties in the food SC, while the ecological emphasises the need to minimize the environmental impact of food production and distribution. Finally, the social dimension underscores the importance of fostering food systems that are equitable, inclusive, and culturally respectful. Thus, sustainable consumption is characterized by "*the utilization of goods and services aimed at diminishing environmental impacts, encouraging social equity, and bolstering economic prosperity both presently and for future generations*" (UNEP, 2017) while considered an essential cornerstone of sustainable development (Wang et al., 2019). Sustainable food consumption is increasingly gaining momentum as individuals and societies become more aware of the environmental, social, and health implications of their dietary patterns (Quoquab et al., 2019).

Sustainable food consumption practices necessitate a thorough comprehension of consumer behaviour, as consumers evaluate market product attributes and make purchasing decisions that can significantly influence their consumption patterns (Hsu et al., 2021). In order to investigate their intricate interrelationships with sustainable food choices, several elements that impact more significant green behaviours, such as needs, interests, and triggers that influence sustainable food consumption, were explored through a systematic analysis.

When considering food choices, the concept of "need" serves as both the foundation and motivation for consumer behaviour. While satisfying basic **psychological** needs such as nutrition and energy (Lema-Blanco et al., 2023) is essential, sustainable food consumption also encompasses broader desires, including taste, convenience, cultural preferences, and health benefits. Furthermore, **environmental** awareness has introduced a new dimension of need, where consumers seek products that minimise ecological impact, promoting "eating green" (Holotová et al., 2021). **Economic** considerations are also important, as financial constraints can limit access to sustainable options, often perceived as more costly (Ghaffar & Islam, 2023). **Social** needs further influence food choices, with consumers purchasing products not only for personal nourishment but also to foster social connections and align with the values of their social groups (Salazar et al., 2012). Lastly, core personal values and future time perspectives significantly affect dietary habits, with sustainable food choices often linked to long-term health and ethical considerations (Olsen & Tuu, 2021).

Moving beyond basic needs, consumer interests encompass a broader range of motivations, including personal preferences, ethical considerations, and lifestyle factors, significantly affecting sustainable food consumption. Ethical concerns, such as promoting justice, fair trade, and local economies, drive consumer behaviour towards morally responsible practices, shifting focus from price and quality to ethical aspects (Torán-Pereg et al., 2023). In addition, personal health and wellness are key factors, with many consumers prioritising organic and non-GMO products within the concept of "lifestyles of health and sustainability" (LOHAS). Environmental awareness also plays a critical role, though the link between knowledge and behaviour remains complex (Hsu et al., 2020). Lastly, social media and advertising powerfully shape consumer choices, potentially fostering more sustainable consumption (Simeone & Scarpato, 2020).

The sustainability of food SCs is a priority area of interest, yet little research has been done on how sustainable food consumption could contribute to the food industry and what the most important drivers, barriers, and indicators are for achieving it (Govindan, 2018).

1.3 Materials and Methods

1.3.1 Sample Size and Data Collection

This study utilised an EU-wide survey conducted in October 2023 to explore the influence of consumers' purchasing habits, product types, purchase frequency, and influencing factors on sustainable food consumption. Data collection was facilitated through Prolific, a crowdsourcing platform, ensuring the recruitment of a representative sample of participants across the EU. This method prioritised inclusivity, enabling participation irrespective of geographical location, age, gender, or other demographic characteristics. Participants were remunerated for their time, contributing to a higher response rate.

Post-data collection, an extensive cleaning process was undertaken to eliminate incomplete or disqualified responses. This included filtering out participants who failed to meet specific criteria, such as not passing attention-check questions or completing the survey too quickly. After this process, a final sample size of 3129 EU consumers was retained.

The sample had a mean age of 30.6 years (SD = 9.66) and a varied educational background, with 38.7% holding a Bachelor's degree and 26.4% possessing a Master's degree. Income distribution showed that the majority belonged to the middle-income bracket (60.2%). Gender distribution was nearly equal, with 47.5% male, 46.2% female, and 1.8% identifying as non-binary. Additionally, 85.6% of participants resided in urban or semi-urban areas. The frequencies and other descriptive statistics regarding the demographics of the used sample are inserted in Tables 1.2, 1.3, 1.4, 1.5, and 1.6 in the Annex.

1.3.2 Questionnaire Design

Respondents were presented with a series of questions designed to measure their attitudes towards sustainable food consumption and purchasing, alongside inquiries about their personal beliefs and interests, reflecting the psychological factors examined in this study. The measurement scales for the constructs were adapted from established literature, ensuring the reliability and validity of our metrics.

Initially, participants were asked to express their familiarity with the concept of sustainable behaviour by rating their agreement with five items on a 5-point scale (1 = highly familiar, 5 = not familiar at all). Following this, to evaluate the willingness to pay a premium for sustainable products, participants rated their agreement with the statement, "*I would pay more for a green product that is making efforts to be environmentally sustainable*", using a Likert scale from 1 (strongly agree) to 5 (strongly disagree), based on the work of Yadav and Pathak (2017). Additionally, participants specified the percentage they would be willing to pay extra by choosing from six options (0%, 1–5%, 6–10%, 11–15%, 16–20%, and more than 20%), as adapted from Kang et al. (2012).

Purchase intention was measured using a construct composed of three items, previously validated by Yadav and Pathak (2017). A similar approach was employed to assess purchase behaviour, with participants rating their agreement with three items on a 5-point scale (1 = strongly disagree, 5 = strongly agree). Furthermore, respondents highlighted key factors influencing their purchase of sustainable food products, derived from the research of Verain et al. (2021), and rated their agreement with a series of statements regarding their interests and price acceptability on a 5-point scale (1 = strongly disagree, 5 = strongly agree), as previously applied by Hsu et al. (2020).

In the survey's final section, participants provided detailed demographic information, including gender (Female/Male/Non-binary/Prefer not to disclose), age (categorised into six groups), residential area (Urban/Semi-urban, Rural), educational attainment (divided into six categories with an option for no disclosure), net annual household income (divided into ten categories ranging from 0 to 100,000+ euros, with an option for no disclosure), and the number of earners and financially dependent children and adults in the household. An attention-check question was incorporated to ensure the quality of responses.

1.4 Results and Discussion

1.4.1 Familiarity Assessment

First, we performed a cluster analysis to determine the sample's familiarity with the concept of sustainable behaviour based on their annual net household income and educational attainment (Fig. 1.1). The cluster analysis diagram demonstrates a

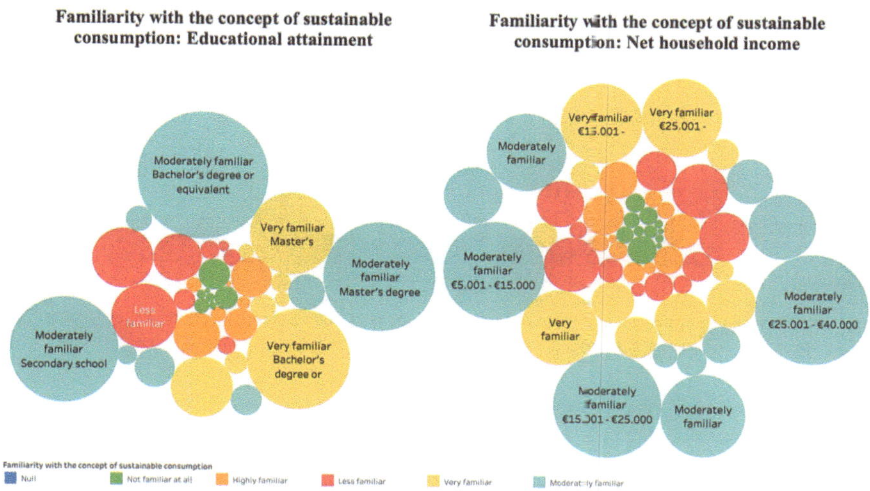

Fig. 1.1 Cluster analysis. (Generated by Tableau 2024.2.2)

moderate and higher familiarity of the sample regarding the concept of sustainable consumption. Firstly, the diagram highlights a strong correlation between educational attainment and familiarity with sustainable consumption. Individuals with higher levels of education, such as those with a bachelor's degree or higher (n = 1.848), are more likely to be familiar with the concept. Higher education may provide individuals with the knowledge and skills necessary to engage with complex issues like sustainability, enabling them to recognize the importance of sustainable practices. Secondly, Fig. 1.1 demonstrates a connection between net household income and familiarity with sustainable consumption. Respondents with lower incomes (< €40.000) are more likely to be familiar with the concept (n = 1.719). The latter observation indicates that income does not act as a barrier to access to sustainability-related information and resources, such as educational materials, environmentally friendly products, and sustainable practices. Despite having less disposable income to invest in sustainable choices, their exposure to and familiarity with the concept is evident.

1.4.2 *Willingness to Pay a Premium*

Figure 1.2 presents an individual's willingness to pay a premium for sustainable products based on their net household income. The data suggests that individuals are willing to pay a premium for sustainable products regardless of net household income. Specifically, the analysis results indicate that about 50–55% of consumers, regardless of net household income, agree to pay a premium for culinary products

Willingness to pay premium: I would pay more for a green product that is making efforts to be environmentally sustainable

Net Household Income

Willingness to pay premium: Green product efforts	Null	€5.000 or less	€5.001 - €15.000	€15.001 - €25.000	€25.001 - €40.000	€40.001 - €55.000	€55.001 - €70.000	€70.001 - €85.000	€85.001 - €100.000	€100.001 or more	I would rather not disclose it
Strongly disagree	3.41%	5.83%	4.98%	2.26%	3.39%	3.96%	3.49%	2.78%	3.33%	2.44%	4.17%
Disagree	15.91%	16.59%	18.33%	14.38%	12.74%	14.52%	14.53%	11.11%	13.33%	12.20%	16.67%
Neither agree nor disagree	29.55%	25.11%	22.95%	21.65%	24.35%	22.11%	16.28%	18.06%	16.67%	26.83%	27.38%
Agree	48.86%	47.98%	50.00%	54.93%	55.48%	53.80%	59.88%	59.72%	61.67%	51.22%	48.81%
Strongly agree	2.27%	4.48%	3.74%	6.79%	4.03%	5.61%	5.81%	8.33%	5.00%	7.32%	2.98%

Willingness to pay premium: I would be willing to pay this extra percentage on the sustainable products to support the organization's/product efforts to be environmentally sustainable

Net Household Income

Willingness to pay premium: Pay an extra percentage	Null	€5.000 or less	€5.001 - €15.000	€15.001 - €25.000	€25.001 - €40.000	€40.001 - €55.000	€55.001 - €70.000	€70.001 - €85.000	€85.001 - €100.000	€100.001 or more	I would rather not disclose it
0%	9.09%	7.62%	6.23%	5.17%	6.13%	7.92%	5.81%	8.33%	8.33%	9.76%	11.31%
1 - 5%	37.50%	40.81%	39.86%	37.96%	40.16%	34.98%	30.23%	25.00%	36.67%	34.15%	33.93%
6 - 10%	32.95%	31.39%	33.81%	32.63%	35.65%	37.29%	38.37%	43.06%	20.00%	29.27%	34.23%
11 - 15%	18.18%	9.87%	13.88%	16.16%	12.90%	12.54%	19.77%	13.89%	23.33%	19.51%	13.69%
16 - 20%	1.14%	8.52%	5.16%	6.14%	4.03%	5.94%	4.07%	6.94%	11.67%	4.88%	6.25%
More than 20%	1.14%	1.79%	1.07%	1.94%	1.13%	1.32%	1.74%	2.78%		2.44%	0.60%

Fig. 1.2 Willingness to reimburse a price premium and premium level (percentages calculated per net household income level). (Generated by Tableau 2024.2.2)

that are supplied sustainably. This observation indicates that Europeans are aware of sustainable food consumption.

Regarding the premium level, the responses indicate that, on average, consumers are willing to pay up to a 10% premium for sustainably sourced food products. For consumers with higher net household incomes (> €40,000), higher price premiums (up to 15%) seem to be acceptable as they may have more disposable income to allocate towards environmentally friendly products.

1.4.3 Purchase Intention

Regarding the purchasing intention, Fig. 1.3a suggests a correlation between higher net household income and a greater likelihood of purchasing sustainable products. Similarly, Fig. 1.3b suggests that individuals with higher net household incomes are more likely to make an effort to purchase sustainable products. Finally, Fig. 1.3c captures the same trend in the willingness to purchase sustainable food products.

In this case, the chi-square test (X^2) of independence of two qualitative variables was applied. The Crammer's V correlation coefficient was used to measure the

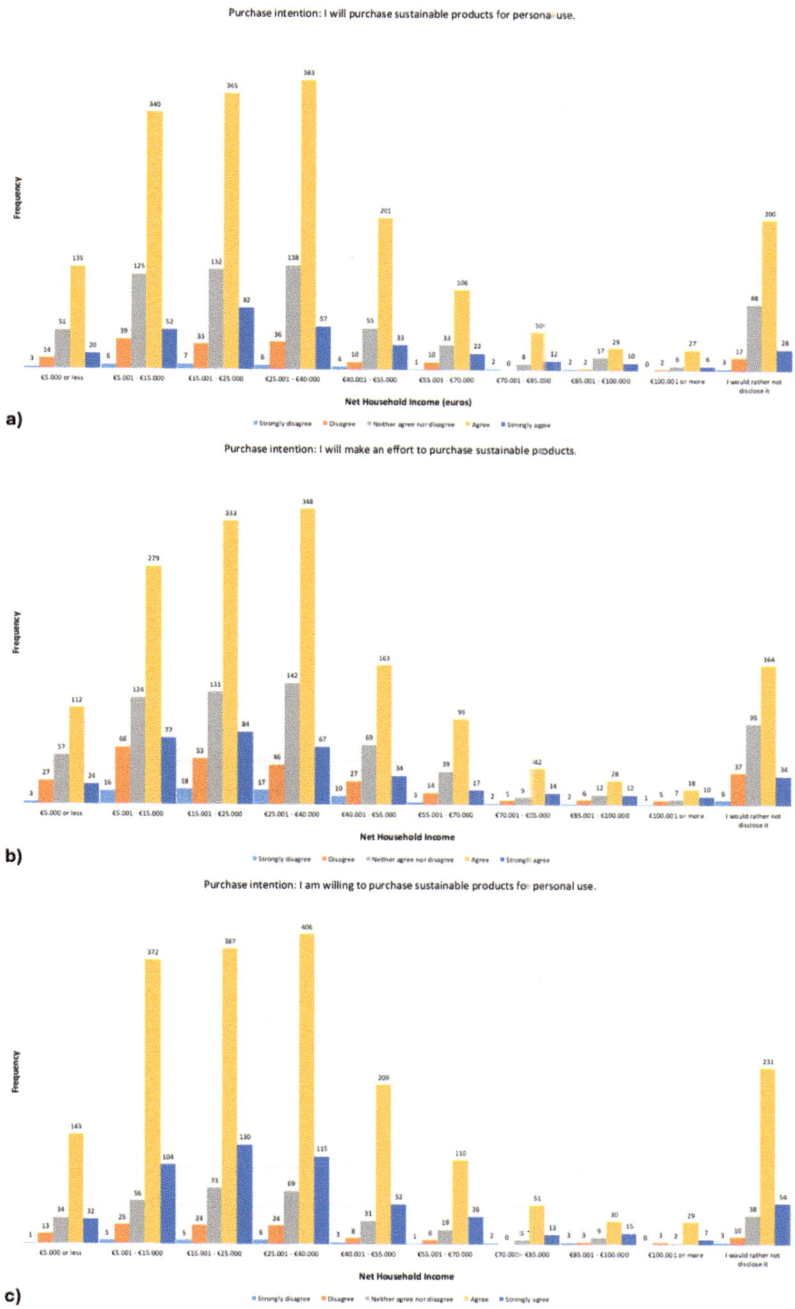

Fig. 1.3 Purchasing intention

Table 1.1 Observed significance of the X^2 test and value of Crammer's V correlation coefficient regarding the possible dependence between the net household income and the degree of purchasing intentions across components of sustainable food products

Variable	Significance (p-value)	Crammer's V
Purchase intention: I will purchase sustainable products for personal use	0.000	0.237
Purchase intention: I am willing to purchase sustainable products for personal use	0.000	0.237
Purchase intention: I will make an effort to purchase sustainable products	0.000	0.238

degree of relevance. The test was carried out at a 5% level of significance. The observed significance of the test conducted in each case and the measure of the coefficient is presented in Table 1.1. The results show a correlation between respondents' net household income and purchasing intention toward sustainable food products, which is of moderate strength (Crammer's V is around 0.237).

1.5 Conclusions

Sustained efforts are essential to developing innovative, sustainable food products that are cost-effective and appealing to consumers. By reducing costs and improving the efficiency of the food supply chain for sustainable options, businesses can broaden their market reach and promote positive change. Companies can customize their strategies to effectively market sustainable food products based on different income segments. For example, highlighting the environmental benefits and social impact of sustainable choices may resonate more with higher-income consumers, while focusing on affordability and accessibility may be more effective for lower-income consumers.

This study explored the relationship between sustainable consumption and supply chain management, particularly within the European Union. Key findings indicate a strong correlation between higher education levels and awareness of sustainable consumption, suggesting that education is vital in promoting sustainable practices. Additionally, the research revealed that consumers across all income levels are willing to pay a premium for sustainable products, with a significant proportion willing to pay up to a 10% premium. This willingness to invest in sustainable options reflects a growing awareness and concern for environmental and social responsibility among consumers.

Furthermore, the study found a positive association between higher household income and the likelihood of purchasing sustainable products, indicating that income can influence sustainable consumption patterns. These findings highlight the importance of considering educational and economic factors in shaping sustainable consumption behaviours and suggest that businesses and policymakers can leverage these insights to promote sustainable practices and drive positive change in supply chain management.

The findings of this study have significant implications for management practices. The strong link between higher education and awareness of sustainable consumption suggests that businesses should invest in initiatives that promote sustainable products and practices. Collaborating with educational institutions, NGOs, or government agencies could help develop effective campaigns. Additionally, consumers across all income levels are willing to pay a premium for sustainable products, indicating a strong market. Managers should make sustainability a core value and explore pricing strategies that accommodate various income segments, such as offering diverse price points or incentives for bulk purchases. Moreover, targeting higher-income consumers, who are more likely to purchase sustainable products, may enhance marketing efforts. This could involve creating premium product lines or exclusive membership programs that offer extra benefits to environmentally conscious buyers.

The present study provides valuable insights into sustainable consumption and supply chain management, but it has limitations. It mainly focused on consumer perspectives and did not adequately address the challenges businesses face in adopting sustainable practices. Future research should examine these barriers and enablers from a business viewpoint, considering factors like cost constraints, technology, and regulations. The reliance on self-reported data may also bias the findings regarding consumer behaviour. More objective measures, such as purchasing data or experimental interventions, could be beneficial. Finally, since the study focused on the European Union, its findings may not apply universally. Future research should investigate sustainable consumption and supply chain management across different regions to identify specific challenges and opportunities.FundingThis research was funded by the ECO-READY project, European Union's HORIZON-CL6–2022 research and innovation programme, grant agreement No. 10108420.

Annex

Table 1.2 Gender frequency data analysis

	Frequency	Percent	Cumulative percent
Null	121	3.87%	0.04
Male	1487	47.52%	51.39%
Female	1447	46.24%	97.64%
Non-binary	57	1.82%	99.46%
I would rather not say	17	0.54%	100.00%
Total	**3129**	**100.00%**	

Table 1.3 Age descriptive statistics analysis

Statistical Measure		Value
Mean		30.601
Median		28
Mode		24
Std. deviation		9.6632
Variance		93.377
Percentiles	25%	24
	50%	28
	75%	35

Table 1.4 Educational attainment frequency data analysis

	Frequency	Percent	Cumulative percent
Null	121	3.87%	3.87%
Primary school diploma	27	0.86%	4.73%
Secondary school diploma	747	23.87%	28.60%
Bachelor's degree or equivalent	1212	38.73%	67.34%
Master's degree	825	26.37%	93.70%
Doctorate's degree (PhD)	83	2.65%	96.36%
Other type of diploma (e.g., professional programmes)	69	2.21%	98.56%
I would rather not disclose it	45	1.44%	100.00%
Total	**3129**	**100.00%**	

Table 1.5 Net household income frequency data analysis

	Frequency	Percent	Cumulative percent
Null	121	3.87%	0.04
€5.000 or less	223	7.13%	0.11
€5001–€15.000	562	17.96%	0.29
€15,001–€25.000	619	19.78%	0.49
€25,001–€40.000	620	19.81%	0.69
€40,001–€55.000	303	9.68%	0.78
€55,001–€70.000	172	5.50%	0.84
€70,001–€85.000	72	2.30%	0.86
€85,001–€100.000	60	1.92%	0.88
€100,001 or more	41	1.31%	0.89
I would rather not disclose it	336	10.74%	1.00
Total	**3129**	**100.00%**	

Table 1.6 Earners' frequency data analysis

	Frequency	Percent	Cumulative percent
Null	169	5.40%	5.40%
1	824	26.33%	31.74%
2	1384	44.23%	75.97%
3	463	14.80%	90.76%
4	189	6.04%	96.80%
>4	60	1.92%	98.72%
None	40	1.28%	100.00%
Total	**3129**	**100.00%**	

References

Anastasiadis, F., Manikas, I., Apostolidou, I., & Wahbeh, S. (2022). The role of traceability in end-to-end circular agri-food supply chains. *Industrial Marketing Management, 104*, 196–211. https://doi.org/10.1016/j.indmarman.2022.04.021

Anastasiadis, F., Apostolidou, I., & Tsolakis, N. (2024). Challenges and opportunities of supply chain traceability: Insights from emergent agri-food sector. *Supply Chain Management: An International Journal*. https://doi.org/10.1108/SCM-05-2024-0322

Carter, C. R., & Rogers, D. S. (2008). A framework of sustainable supply chain management: Moving toward new theory. *International Journal of Physical Distribution and Logistics Management, 38*(5), 360–387. https://doi.org/10.1108/09600030810882816

Cooper, M. C., Lambert, D. M., & Pagh, J. D. (1997). Supply chain management: More than a new name for logistics. *The International Journal of Logistics Management, 8*(1), 1–14. https://doi.org/10.1108/09574099710805556

De Oliveira, U. R., Gomes, T. S. M., De Oliveira, G. G., De Abreu, J. C. A., Oliveira, M. A., Da Silva César, A., & Aprigliano Fernandes, V. (2022). Systematic literature review on sustainable consumption from the perspective of companies, people and public policies. *Sustainability, 14*(21), 13771. https://doi.org/10.3390/su142113771

Folkerts, H., & Koehorst, H. (1997). Challenges in international food supply chains: Vertical coordination in the European agribusiness and food industries. *Supply Chain Management: An International Journal, 2*(1), 11–14. https://doi.org/10.1108/13598549710156312

Ghaffar, A., & Islam, T. (2023). Factors leading to sustainable consumption behavior: An empirical investigation among millennial consumers. *Kybernetes, 53*(8), 2574–2592. https://doi.org/10.1108/K-12-2022-1675

Govindan, K. (2018). Sustainable consumption and production in the food supply chain: A conceptual framework. *International Journal of Production Economics, 195*, 419–431. https://doi.org/10.1016/j.ijpe.2017.03.003

Holotová, M., Horská, E., & Nagyová, Ĺ. (2021). Changing patterns of sustainable food consumption regarding environmental and social impact-insights from Slovakia. *Frontiers in Sustainable Food Systems, 5*. https://doi.org/10.3389/fsufs.2021.703827

Hsu, S.-Y., Wang, H.-C., Ho, J.-L., & Chen, H.-C. (2020). Exploring consumers' interest in choosing sustainable food. *Frontiers in Psychology, 11*, 489. https://doi.org/10.3389/fpsyg.2020.00489

Hsu, H.-W., Chen, C.-Y., & Wu, C.-W. (2021). Cross-cultural comparison of sustainable agro-food consumption from consumers' perspectives: Cases from Taiwan and France. *Sustainability, 13*(17), 9733. https://doi.org/10.3390/su13179733

Hu, J., Liu, Y.-L., Yuen, T. W. W., Lim, M. K., & Hu, J. (2019). Do green practices really attract customers? The sharing economy from the sustainable supply chain management perspective. *Resources, Conservation and Recycling, 149*, 177–187. https://doi.org/10.1016/j.resconrec.2019.05.042

Kang, K. H., Stein, L., Heo, C. Y., & Lee, S. (2012). Consumers' willingness to pay for green initiatives of the hotel industry. *International Journal of Hospitality Management, 31*(2), 564–572. https://doi.org/10.1016/j.ijhm.2011.08.001

Lema-Blanco, I., García-Mira, R., & Muñoz-Cantero, J.-M. (2023). Understanding motivations for individual and collective sustainable food consumption: A case study of the Galician conscious and responsible consumption network. *Sustainability, 15*(5), 4111. https://doi.org/10.3390/su15054111

Lim, L. L. K., & Londob, G. (2018). Global consumer cultures and experiences of sustainable food consumption in host country. *Journal of Applied Business and Economics, 20*(4). https://doi.org/10.33423/jabe.v20i4.346

Linton, J. D., Klassen, R., & Jayaraman, V. (2007). Sustainable supply chains: An introduction. *Journal of Operations Management, 25*(6), 1075–1082. https://doi.org/10.1016/j.jom.2007.01.012

Mentzer, J. T., DeWitt, W., Keebler, J. S., Min, S., Nix, N. W., Smith, C. D., & Zacharia, Z. G. (2001). Defining supply chain management. *Journal of Business Logistics, 22*(2), 1–25. https://doi.org/10.1002/j.2158-1592.2001.tb00001.x

Mitlin, D. (1992). Sustainable development: A guide to the literature. *Environment and Urbanization, 4*, 111–124. https://doi.org/10.1177/095624789200400112

Olsen, S. O., & Tuu, H. H. (2021). The relationships between core values, food-specific future time perspective and sustainable food consumption. *Sustainable Production and Consumption, 26*, 469–479. https://doi.org/10.1016/j.spc.2020.12.019

Poore, J., & Nemecek, T. (2018). Reducing food's environmental impacts through producers and consumers. *Science, 360*(6392), 987–992. https://doi.org/10.1126/science.aaq0216

Quoquab, F., Mohammad, J., & Sukari, N. N. (2019). A multiple-item scale for measuring "sustainable consumption behaviour" construct: Development and psychometric evaluation. *Asia Pacific Journal of Marketing and Logistics, 31*(4), 791–816. https://doi.org/10.1108/APJML-02-2018-0047

Salazar, M. R., Lant, T. K., Fiore, S. M., & Salas, E. (2012). Facilitating innovation in diverse science teams through integrative capacity. *Small Group Research, 43*(5), 527–558. https://doi.org/10.1177/1046496412453622

Sánchez-Flores, R. B., Cruz-Sotelo, S. E., Ojeda-Benitez, S., & Ramírez-Barreto, M. E. (2020). Sustainable supply chain management – A literature review on emerging economies. *Sustainability, 12*(17), 6972. https://doi.org/10.3390/su12176972

Sarkis, J., Zhu, Q., & Lai, K. (2011). An organizational theoretic review of green supply chain management literature. *International Journal of Production Economics, 130*(1), 1–15. https://doi.org/10.1016/j.ijpe.2010.11.010

Seuring, S., & Müller, M. (2008). From a literature review to a conceptual framework for sustainable supply chain management. *Journal of Cleaner Production, 16*(15), 1699–1710. https://doi.org/10.1016/j.jclepro.2008.04.020

Simeone, M., & Scarpato, D. (2020). Sustainable consumption: How does social media affect food choices? *Journal of Cleaner Production, 277*, 124036. https://doi.org/10.1016/j.jclepro.2020.124036

Torán-Pereg, P., Mora, M., Thomsen, M., Palkova, Z., Novoa, S., & Vázquez-Araújo, L. (2023). Understanding food sustainability from a consumer perspective: A cross-cultural exploration. *International Journal of Gastronomy and Food Science, 31*, 100646. https://doi.org/10.1016/j.ijgfs.2022.100646

UN DESA. (2023). *The sustainable development goals report 2023: Special edition*. https://unstats.un.org/sdgs/report/2023/

UNEP, U.N. (2017, October 2) Sustainable consumption and production policies | UNEP – UN Environment Programme. https://www.unep.org/explore-topics/resource-efficiency/what-we-do/sustainable-consumption-and-production-policies. Accessed 14 Aug 2024.

Verain, M. C. D., Snoek, H. M., Onwezen, M. C., Reinders, M. J., & Bouwman, E. P. (2021). Sustainable food choice motives: The development and cross-country validation of the sustainable food choice questionnaire (SUS-FCQ). *Food Quality and Preference, 93*, 104267. https://doi.org/10.1016/j.foodqual.2021.104267

Wang, C., Ghadimi, P., Lim, M. K., & Tseng, M.-L. (2019). A literature review of sustainable consumption and production: A comparative analysis in developed and developing economies. *Journal of Cleaner Production, 206*, 741–754. https://doi.org/10.1016/j.jclepro.2018.09.172

Yadav, R., & Pathak, G. S. (2017). Determinants of consumers' green purchase behavior in a developing nation: Applying and extending the theory of planned behavior. *Ecological Economics, 134*, 114–122. https://doi.org/10.1016/j.ecolecon.2016.12.019

Zhang, T. C., Gu, H., & Jahromi, M. F. (2019). What makes the sharing economy successful? An empirical examination of competitive customer value propositions. *Computers in Human Behavior, 95*, 275–283. https://doi.org/10.1016/j.chb.2018.03.019

Chapter 2
Collaboration, Resilience, and Value Creation in the Scottish Pig Supply Chain

Cesar Revoredo-Giha, Wisdom Dogbe, and Luiza Toma

2.1 Introduction

The pig industry in Scotland is relatively small compared to the cattle and sheep sectors. According to Quality Meat Scotland (QMS, 2023), in 2020, the red meat sector (i.e., cattle, sheep, and pig production) contributed £1.29 billion to the total agricultural output (i.e., 37%), of which pig production was 14.8% (£139 million) and 4% of the total agricultural output.

The pig sector has not historically benefited directly from government support (i.e., Single Farm Payment) through the European Union's Common Agricultural Policy. This has made the situation of the sector more challenging as it has been exposed to significant shocks. Due to the disruptions from the COVID-19 pandemic, the sector has been operating at a substantial and growing loss (Scottish Government, 2021). Information from The Scottish Farmer (2022) indicated that an East Lothian pig farmer estimated that, while by February 2021 the sector was experiencing losses of around £26 per pig, these losses increased by March 2022 to £56 per pig. This situation was explained by the influx of pork imports by retailers and the rise in feed prices resulting due to the Russian-Ukraine situation.

The development of the Scottish pig supply chain and its reaction to shocks are interesting to study because on the one hand, it has many features in common with other agri-food supply chains, particularly meat ones, and on the other hand, it allows us to illustrate how, particularly horizontal, collaboration in the form of pig cooperatives have played a role to reduce the vulnerability of producers, i.e.,

C. Revoredo-Giha (✉) · L. Toma
Scotland's Rural College (SRUC), King's Buildings, Edinburgh, UK
e-mail: cesar.revoredo@sruc.ac.uk; luiza.toma@sruc.ac.uk

W. Dogbe
The Rowett Institute, University of Aberdeen, Aberdeen, UK
e-mail: wisdom.dogbe@abdn.ac.uk

K. Mattas et al. (eds.), *Biodiversity in Agri-Food Systems*, Cooperative Management, https://doi.org/10.1007/978-3-032-06385-4_2

increase their resilience to several shocks, and also on their process of value creation. Moreover, although the sector has sporadically established vertical collaboration arrangements with its customers, these have been more difficult to achieve (Leat & Revoredo-Giha, 2013a, b).

The purpose of this paper is to highlight how changes on the business environment where supply chains operate affect their performance in terms of resilience and value creation and also their reaction to the changes. For this, it is necessary to consider a historical perspective as many of the variables within the business environment evolve slowly. In addition, the paper aims to discuss the potential consequences that the requirements for the UK net zero strategy by 2050 may have for the Scottish pig supply chain.

The structure of the paper is as follows: it starts with a brief literature review of the main concepts of the paper, namely: risk and resilience, collaboration, and value creation. The next section introduces the Scottish pig supply chain and its evolution, highlighting the trends in major variables affecting the chain. Then, the paper discusses the role of collaboration on increasing resilience, generating value creation, and how changes in the structure of the Scottish supply chain and the UK environmental policy can affect the vertical relationships of the chain. The last section presents the final remarks.

2.2 Literature Review

This literature review covers the topics of risk and resilience, collaboration and value creation in food supply chains.

2.2.1 Risk and Resilience

Bodie and Merton (2000) describe risk as an uncertain event that affects the welfare of an individual or business and is often associated with adversity or loss. Furthermore, Mason-Jones and Towill (1998) categorized risk into two broad themes: (1) internal risk which is made up of both process and control risks and (2) external risks comprising of demand (downstream) and supply (upstream) risk factors. Note that the concept of risk is related to vulnerability, and supply chain vulnerability involves the identification of risks which affect not only the individual agent but also the entire supply chain (Leat & Revoredo-Giha, 2013a).

The second concept related to risk is that of resilience. Focusing on the business level, Briano et al. (2009) noted that resilience refers to the ability of a company to resist a serious damaging event. Related to the latter, Christopher and Peck (2004) defined resilience as the ability of a system to return to its original state or move to a new, more desirable state after being disturbed, while Fiksel (2003) suggested that resilient systems satisfy four major characteristics, namely, diversity, efficiency, adaptability and cohesion.

At the supply chain level, resilience aims at developing the adaptive capability of the chain to prepare for unexpected events and to respond to disruptions and recover from them (Ponomarov & Holcomb, 2009). Thus, a company or supply chain that has developed its resilience is capable to support the unpredictability on its operation, thereby achieving a competitive advantage through being able to recover more quickly than competitors when an adverse event arises.

Note that implicit in the development of chain resilience is the process of risk management, which is a continuous cycle of identification of hazards, assessment of risks, analysis and selection of controls, implementation and review (Manuele, 2005). Pettit (2008) brought these ideas together to form a conceptual framework which involves the vulnerability of a business or chain to various forms of external and internal hazards or risks and its capability to cope with these. In this context, capability is regarded as arising from a series of factors such as flexibility, spare capacity, efficiency, visibility, adaptability, anticipation, recovery, asset dispersion, collaboration, organisation, market position, security and financial strength. Pettit also identified a zone of balanced resilience in which an enterprise or chain's abilities to cope with vulnerability to risk are balanced by its capabilities to cope with them.

2.2.2 *Collaboration*

Although there are several views on how to define supply chain collaboration (Barratt, 2004), some common features are typical. Thus, collaboration involves multiple firms or autonomous business entities engaging in a relationship that aims to share improved outcomes and benefits. In addition, collaboration can manifest in a form of vertical collaboration (a relationship among players from upstream to downstream along the supply chain) and horizontal collaboration (a relationship among stakeholders that play at the same level) (Dania et al., 2016).

According to Dania et al. (2016) collaboration has the obvious advantages of minimising cost, increasing profit margins, and fulfilling quality assurance and the trust of customers. To achieve these improvements in performance, the business entities need to establish an appropriate level of trust; share critical information; make joint decisions; and, when necessary, integrate supply chain processes. Moreover, collaboration is a key strategy regarding the management of risks as firms are affected by numerous supply chain disruptions; as a result, working with supply chain partners and stakeholders is imperative to tackling these disruptions.

It is important to highlight that the intensity of collaboration among actors is influenced by the level of motivation, trust and the willingness to share risks among the supply chain participants (Nicolaas Bezuidenhout et al., 2012; Steele & Feyerherm, 2014).

2.2.3 Value Creation

Value creation is a prime component of the operation of supply chains. Value is created as a product moves from one stage of the chain to another. It can be defined as the process whereby the capabilities of supply chain actors are joined together such that the competitive advantage of the entire supply chain is improved (Borys & Jemison, 1989).

Following Insch (2008), value can be categorised into two broad types: value in use and value in exchange. The former relates to the utility of a particular object or good, whereas the latter relates to its power to purchase other goods in exchange (Smith, 1999).

Value in exchange is concerned about the monetary or tangible value added to the product or service at each stage of the production process, i.e., from raw material to final disposal. From the accounting perspective, the value of a product or service is expressed as the "selling price minus the cost of raw materials and production activities" (Insch, 2008).

The value in the use of a product deals with the total utility (tangible and intangible) derived from consuming a product or using a service. The value in use is derived from product attributes that provide utility to the end-user—similar to Lancaster's hedonic theory (Lancaster, 1966). Utility to the end-user concerns both functional and psychological benefits derived from consuming a product or using a service.

In the business context, the creation of value links suppliers, firms, and customers, defining the concept as the difference between what consumers are willing to pay for and suppliers' opportunity cost (Brandenburger & Stuart, 1996; Helfat et al., 2009). McIntyre et al. (2022) explained that value creation within the value chain relies on understanding the end user, products offering attributes, the production process as well as communication between supply actors.

2.3 Research Approach

The aim of this study is to analyse the evolution of the Scottish pig supply chain and its efforts to achieve resilience to changes in the market environment. As proposed by Insch (2008), a historical research approach is the most appropriate one in these cases since description and analysis of previous events and conditions provide insights into the evolution of the activity.

The above methodological approach also allows the researcher to use multiple data sources to examine events in their full context and apply a range of techniques for analysis and synthesis. This helps to construct a historical narrative coming from different collated sources. This is particularly important given that the size of the chain may imply that there was limited information.

The sources used in this work came from three categories: (1) statistical compendiums and documents, which comprise government parliamentary records, policy documents, organisational records and statistical yearbooks (e.g., Scotland Agricultural Statistics); (2) processing of surveys, which relates to the processing of the National Food Survey and the Living Costs and Expenditure Survey covering the period 1974–2021 in total; and (3) newspaper information.

2.4 Evolution of Selected Factors Affecting the Scottish Pig Chain

This section analyses the evolution of several factors that have had an influence on the pig supply chain. These factors include: the evolution of the pig supply, the development of pig meat processors, wholesale meat traders' investment in abattoirs, retailers' evolution in negotiation power, and consumer trends regarding pig products.

2.4.1 Pig Supply

Figures 2.1 and 2.2 present two components of the pig supply: the pig stocks and imports of major pig products. Whilst Fig. 2.1 refers to Scotland, Fig. 2.2 is only available for the UK; this is due to the fact that imports occur at the level of the country (i.e., UK) and they are distributed to each one of the UK nations.

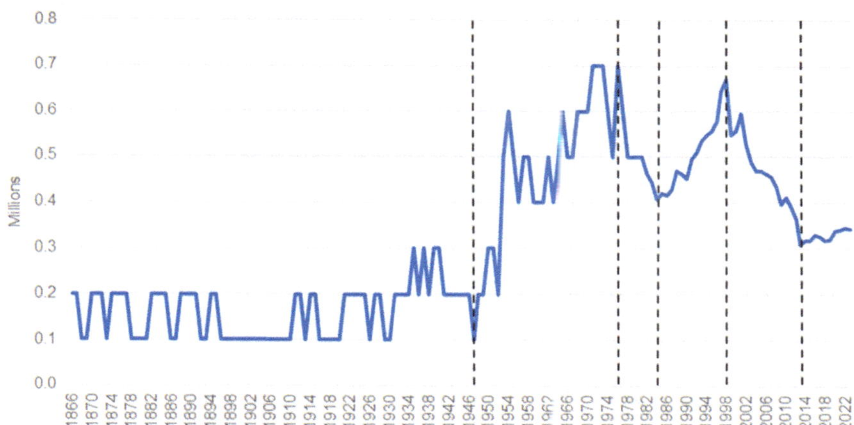

Fig. 2.1 Scotland—Pig stocks 1866–2023. (Source: Agricultural Statistics—Scotland, Economic Report on Scottish Agriculture, various numbers)

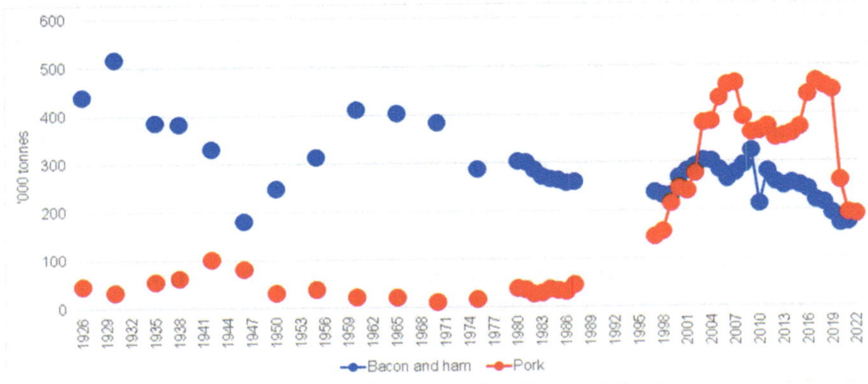

Fig. 2.2 United Kingdom—Pork, bacon and ham imports 1926–2022. (Source: Marks and Britton (1989), Custom and Excise and Defra)

A long-term series of animals finished (i.e., slaughtered) in Scotland is not available. However, pig stocks in Scotland are available since 1866; this is useful because they are the production basis (i.e., finished animals). A change in the pig stocks is a leading indicator of the production of pigs.

Figure 2.1 shows that the stocks of pigs remained stable from 1866 to the end of the Second World War. Despite ups and downs, stocks triplicated from 1946 to 1978. This was followed by a sharp decrease between 1978 and 1984, followed by a recovery that reached its maximum in 1998. From 1998 to about 2014, the stock of pigs suffered another significant decrease, from approximately 0.7 million pigs to about 0.3 million pigs, a figure not seen since the 1950s. Since 2014, the trend has been positive but less dramatic than in the past.

Series of the imports of major pig products (i.e., pork, bacon and ham) are available only since 1926 (Fig. 2.2). Until the late 1990s, the imports of bacon and ham were above those of pork. Whilst the imports of bacon broadly remained at historically similar levels (e.g., around 280 million tonnes), imported pork on average increased about 8 times since 1996 with respect to its previous values. The increase in imports coincided with a contraction of the stock of pigs in Scotland.

2.4.2 *Expansion of Abattoirs*

A key stage in the development of livestock supply chains is the construction of abattoirs and further processing. There are no historical statistics about the number of pig abattoirs in Scotland before 1996; therefore, the historical information comes from sources such as newspapers like The Press and Journal, which is a daily regional newspaper serving Northern and Highland Scotland. The expansion of abattoirs in Scotland came from two main sources: (1) expansion of auction marts into the wholesale meat market, and (2) the development of the pig meat processing industry. This section concentrates on large abattoirs present before 1980.

2.4.2.1 Auction Marts Investment in the Wholesale Meat Market

As mentioned by The Press and Journal (24-02-1976), the story of Aberdeen and Northern Marts (known as "the Marts") goes back to 1872 when a group of farmers got together and were selling as the Aberdeen Cattle and Farm Produce Association. By 1975, the turnover had grown to £6 million, of which approximately half was made of operations in Scotland, a third in England and a sixth abroad.

It was not until 1953 that the Aberdeen and Northern Marts started to think about the wholesale meat business (i.e., abattoirs), which coincided with the last years of the end of the UK Ministry of Food (1939–1958). However, it did not materialise until 1963 when an abattoir at Banchory (Aberdeenshire) was built and the Aberdeen Meat Marketing Company (AMMC), a subsidiary of the Aberdeen and Northern Marts, was created. More important for the topic of this paper is that three years later, in 1966, an abattoir (mainly for pigs) was built at Buckie (Moray).

In May 1970, Marts inaugurated a new Buckie slaughterhouse which was built close by the old one at the former Buckpool railway station at a cost of £230 thousand and employing a workforce of 40 people (Buckie Heritage, 1970). It should be highlighted that Aberdeen and Northern Marts not only invested in abattoirs for pigs but also in cattle and sheep abattoirs. Thus, by 1975 the Marts invested in the refurbishing of Bathgate (Midlothian), previously owned by the Scottish Meat Producers.

The situation of the Buckie abattoir deteriorated and in July 1985, AMMCO announced that they would close the slaughterhouse (Buckie Heritage, 1985), despite a fight to save the facility for local farmers and the plant jobs, led by the local community, district and regional councillors with backing from the National Farmers Union.

2.4.2.2 Development of Pig Meat Processors

The creation of abattoirs in Scotland was also associated to the development of processors. In this section we concentrate on two of these processors due to their size and importance for the supply chain in Scotland David A. Hall and Lawson's of Dyce.

David A. Hall

The name David Hall or the brand Hall's is associated with the traditional Scottish breakfast. The company supplied sausages and black pudding to Scotland's butchers for several generations.

The origin of Hall's started in 1932 when David A. Hall started a Corstorphine butcher's shop and the company remained family-owned for more than 60 years

(The Scotsman, 2021). Mr. Hall not only saw the business producing sausage, haggis and black pudding quickly take off, but also became honorary treasurer and was elected in 1944 as president of the Edinburgh Master Butchers' Association (Edinburgh Evening News—Thursday 22 February, 1940).

The expansion of the company, however, was due to David A. Hall's sons, David R. and James. David R. took up a butchery apprenticeship in 1948 at his father's shop in the city's Corstorphine district. Working alongside his brother James and father David, David R. expanded the David A. Hall business, moving it to a new facility in Newhouse (Lanarkshire) in the mid-1950s, where they supplied butchers' shops with sausages, bacon and pies, with the business expanding from the initial four vans to a fleet servicing Scotland (The Scotsman, 2021).

The company grew to such an extent that there was a need to move the family firm to the larger site in Broxburn, which was completed in 1962. At its peak the company employed over 1100 people and went on to win a Royal warrant. David R. oversaw expansion again in the late 1980s, gearing it up to service the new supermarket regime, concentrating on supplying the multiple and various food markets in the UK and abroad. Moreover, it was during this period that the company expanded into Woodburn (Dalkeith) but unfortunately the new plant burnt down after three weeks of operation.

In the mid-1990s, the economic situation of the company deteriorated and in April 1996, nearly 100 workers were laid off at the site in the wake of the BSE crisis and fluctuating pig prices. Its time as a family firm ended in 1997, when David R.'s son was replaced as managing director. The company was taken over by Grampian Country Food Group in 1998.

Lawson's of Dyce

The company was founded in 1906 by Robert Lawson when he opened a bacon-curing plant at Dunfermline and was continued by Frank and Robert Lawson, sons of the founder. In 1933, Frank Lawson came across an empty bacon factory in Dyce (Aberdeen) which had been built in 1926 as a farmers' cooperative, but which survived only for a couple of years. When the factory was reopened in 1934, only 30 pigs per week were being slaughtered. The company expanded steadily to become one of the more modern and efficient in Scotland.

Lawson's had more than 900 employees in their organisation and were one of the biggest meat producer companies in Britain. More than 7000 pigs transported from all over Scotland were slaughtered, cured and processed at the Dyce factory each week. The firm produced bacon, sausages, pies, hams, black and white puddings, lard, cooked meats and a number of by-products. It had their own pig breeding farm; in 1964, it installed a £55,000 electronic computer at Dyce (The Press and Journal, 10 July 1965).

Despite its success, in 1965 Lawson's of Dyce was taken over and became a Unilever subsidiary, although it continued trading under the same name. This new arrangement was not to last as pig slaughtering and curing ceased in October 1979, when the company announced the closure of the part of the company related to pig slaughtering and curing, which brought a loss of 600 jobs. The remaining part of the company, which employed 900 people, remained opened, concentrating on developing the manufacture and sale of high-quality sausages, prepacked bacon and cooked meats under the Lawson's brand name.

In 1986, the remaining part of the factory closed, being bought by Walls, and Lawson's became only a label on some products. The company in Dyce who took over Lawson's old factory was MacIntosh of Dyce which supplied pies, sausage rolls and other meat products to supermarket chains until 2002 when the firm went into liquidation.

2.4.3 *Retailers' Evolution and Their Negotiation Power*

According to Hollingsworth (2004), the retailing revolution in the 1970s and 1980s resulted in a number of very significant developments. The first one is that the growth in absolute and relative size of retailers not only replaced the manufacturers' dominance in the supply chain but also eliminated many wholesalers and started the trend towards backward integration of the retailer.

The 1990s witnessed continued concentration and consolidation in all areas of the sector (Davies & Ward, 2000). The four major food retailers in the 1990s (i.e., Tesco, Sainsbury's, Asda and Safeway) dominated grocery retailing. The industry structure became characterised by a number of common attributes including greater store size, increase in retailer concentration and the adoption of a range of formats by retailers to reach as large a customer base as possible.

Burt and Sparks (2003) point out that after the Monopolies and Mergers Commission (MMC) (1981) and Office of Fair Trading (OFT) (1985) studies into competition in the UK retail grocery market, concentration continued to increase. In the UK, the locus of 'power' in the distribution channel shifted away from branded goods manufacturers towards retailers. In many product markets (including grocery), retailers assumed channel leadership, utilising improved demand and customer information to develop their brand position and overall retail offer (Burt & Sparks, 2003; Wileman & Jary, 1997). This evolution can be seen in Fig. 2.3, which shows the increasing share of the total groceries in the hands of Tesco, Asda and Sainsbury's.

As fewer retailers have taken a larger share of the grocery market, the 'gatekeeping' role traditionally assigned to retailers because of their location in the channel has assumed greater significance, raising concerns over the extent and use of retail power (Clarke, 2002). Such concerns had been raised before (e.g., Wrigley, 1987, 1991) but with little impact. In the mid-to-late 1990s, amid widespread press concerns and other reports (e.g., Dobson et al., 1998), the government ordered a Competition Commission investigation of the sector. Whilst the Commission found little evidence of abuse of market power in terms of pricing and profits, it did express concerns over treatment of suppliers.

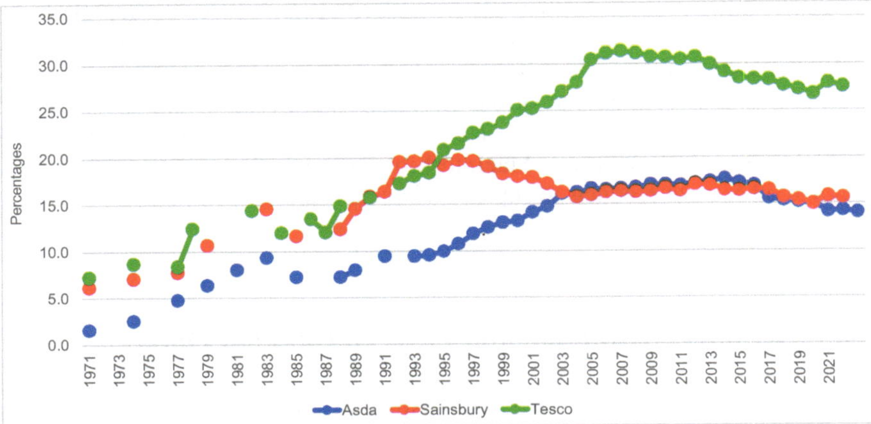

Fig. 2.3 United Kingdom—Major retailers' share on total groceries 1970–2022. (Source: www.fooddessert.org)

2.4.4 Consumer Trends

The presence of a growing demand is important for the thriving of a supply chain. Although total consumption grows due to increases of the population, it is more revealing to consider the evolution of per-capita consumption due to the fact that it provides information about changes in consumers' preferences. Figure 2.4 considers the evolution of per capita purchases of pork, bacon and ham, major products in the pig chain, from 1942 to 2020.

Both series show different starting trends. In the case of pork, the purchases show an increasing trend (with the exception of the post-war period) until about 1980, to start decreasing since then. In the case of bacon and ham, the increase after the war years reached a plateau from the mid-1950s until about the 1970s, to start decreasing since then.

There are several reasons to explain the observed trends: changes in the consumption patterns (particularly breakfast), the increasing competition of other meats (particularly poultry), and trends about health and red meat.

2.5 Supply Chain Situation After 1980

The choice of 1980 to break the story of the pig supply chain is due to the effects that the closure of Lawson's of Dyce abattoir had for the pig producers in Scotland. As a reaction of the end of Lawson's in 1979 and 1980, two pig producer cooperatives were established: Grampian Pig Producers and Scotlean Pigs. Another reason for choosing 1980 is due to the fact that the Scottish pig supply chain entered into a

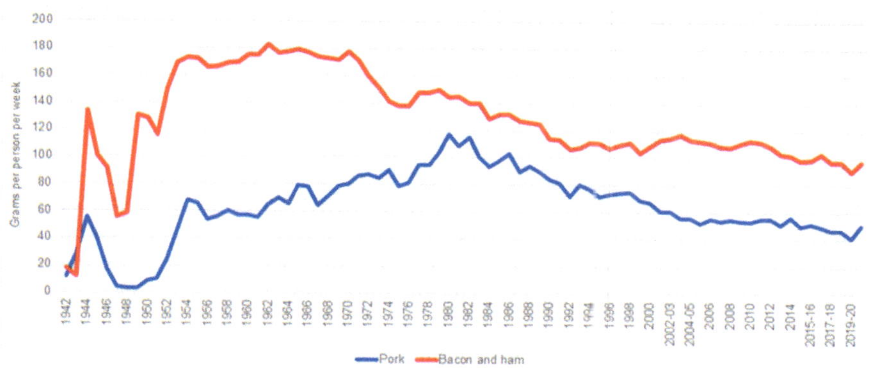

Fig. 2.4 United Kingdom—Pork, bacon and ham consumer trends 1942–2020. (Source: MAFF, National Food Survey, various number; Defra, Family Food)

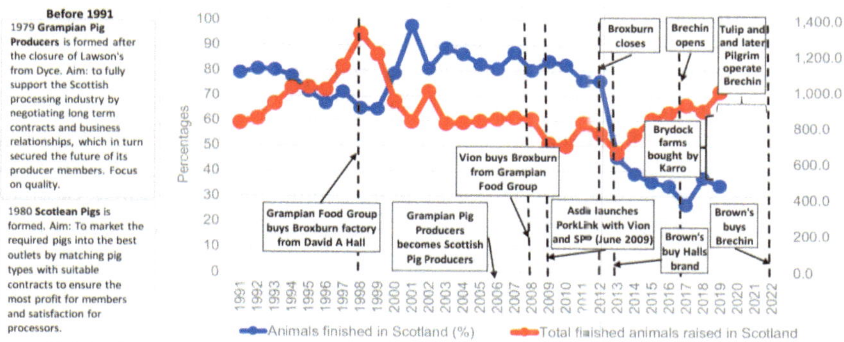

Fig. 2.5 Timeline of major events on the Scottish pig supply chain. (Source: Own elaboration based on ERSA and Scottish Slaughterhouse survey)

period of instability, particularly regarding the situation of the processing part of the chain. Figure 2.5 presents all the events since the 1980s.

2.5.1 *The Formation of the Pig Cooperatives*

2.5.1.1 Scottish Pig Producers

The loss of the processor Lawson's of Dyce in September 1979 led to the formation of Grampian Pig Producers Limited. These were difficult days for producers in the North and especially in the Grampian Region, where some 50% of Scottish pig production was based. The cooperative was very successful and quickly built-up membership and throughput, which stopped the decline in sow numbers and put confidence back in the industry. By 1990, they had grown to 165 members and 7000 pigs per week, and had the full support of producers and processors.

Since its inception, the aim of the cooperative was to fully support the Scottish processing industry by negotiating long-term contracts and business relationships, which in turn secured the future of its producer members.

An important aspect of the cooperative is its focus on quality of service to producer and processor with a track record of pioneering new product lines, working closely with supply chain partners. Scottish Pig Producers also plays a leading role in industry innovations, such as the Wholesome Pigs (Scotland) health scheme, to improve productivity, performance to specification, and ultimately value for farmer members.

Innovations in members' services continued, with the introduction of the immediate payment facility and ISO9001 accreditation. In 2006, in order to better reflect the national membership, Grampian Pig Producers changed its name to Scottish Pig Producers.

Close links were developing with producers in Northern Ireland and were formalised in 2010 when Scottish Pig Producers took on the role of marketing agent for the Progressive Lean Pigs cooperative. This has grown continuously into a substantial part of the overall business, supplying high-quality pigs to the major processing companies operating in Northern Ireland.

Developing the Scottish pig industry has been a key priority for Scottish Pig Producers over the years, with strong links to Scottish Government, Quality Meat Scotland, SAOS, NFU Scotland, Ulster Farmers Union and a wide range of other organisations. Scottish Pig Producers were instrumental in the development of Quality Pork Limited, the collaboration with Scotlean and Tulip to redevelop the Brechin abattoir that can supply top quality 'Specially Selected Pork' to consumers across Scotland and beyond.

2.5.1.2 Scotlean Pigs

Scotlean Pigs was established in Scotland in 1980 by a group of progressive producers who were interested in the potential benefits of cooperation and establishing their own marketing opportunities post-farm gate. The move was prompted by signs of a decrease in the number of slaughter/processors in Scotland, which led to a buyer's market.

The members of the cooperative are situated from Aberdeen to Wiltshire, with a range that goes from family farms to large and intensive businesses. Scotlean's growth developed bespoke and well-structured marketing contracts with new and progressive abattoir/processing companies. This allowed Scotlean to respond to a range of member needs, including the development of competitive purchasing initiatives, which reduce producers' costs without compromising on quality and, ultimately, pig performance.

The cooperative can provide the following benefits to farm businesses: professional 'in-house' pig marketing; a choice of market contracts; assistance in reducing cost of production; provision of a farm recording package; and consultant veterinary expertise for sound pig business profitability. Cooperation can still provide the right opportunities for the British pig producers.

As a result of changes to the size of the UK pig industry that led to increased imports of pig meat, often from countries with lower production costs, Scotlean members have developed an even greater focus on their business processes.

2.5.2 *Evolution of the Processing Sector*

In this section we explore the situation of three largest pig abattoirs in Scotland: Buckie, Broxburn and Brechin.

2.5.2.1 Buckie Abattoir

In October 1985, the Buckie abattoir was sold to Grampian Country Pork, with the expectation that around fifty jobs would be created. In fact, the company invested £1 million to transform the building into one of the most modern pork processing plants in the UK and the workforce grew from the original 40 to about 130. However, the stability lasted only two years as in December 1987, the Buckie Heritage pointed out that recent developments were fuelling speculation over the future of the Grampian Country Pork factory at Buckpool (i.e., Buckie). This was reflected in redundancies and letters being sent by the firm to pig suppliers informing them of an imminent reduction of the price paid for their animals. Moreover, the sales manager, two salesmen, two production managers and a tele-salesperson lost their jobs.

However, in November 1980, the Buckie factory announced expansion plans to increase the capacity to 5000 pigs a week, ensuring the employment of the 100 workers at the plant and creating 15 more employment positions in the areas of butchering and packing. The plant, then, was the only one satisfying the requirements of the European Economic Community in Scotland, starting exports to Germany, Ireland and at a minor measure to France.

The abattoir also sold to the domestic market of which about 50% was to supermarkets, Tesco being the biggest customer, followed by Gateway, William Low, The Co-op and Safeway. The other 50% went to manufacturers of ham, pies and sausages.

On March 2005, Grampian Country Pork decided to close the factory in Buckie with consequences for the workers who relied on the factory for their livelihoods, and for the economy of the town. By April, the MP for Moray and senior managers of Grampian Country Foods Groups maintained the conversation to try to find a solution. However, seven days later, the company announced that the factory was to close its doors on October with the loss of around 330 jobs.

2.5.2.2 Broxburn Abattoir

In parallel to the events related to the Buckie abattoir, in April 2005, the Grampian Country Food Group (owner of Grampian Country Pork) announced a £16 million re-development and expansion of the pork processing business at Broxburn, near Edinburgh.

Grampian Country Food Group comprised farms, processors and supplies for chicken, turkey, beef, lamb and pork products to most British supermarkets from 30 principal plants in the UK and Thailand. During that decade, they moved into pork and agricultural feed, working from new facilities in the Scottish central belt and England and Wales. They took on brands like Marshalls of Newbridge, makers of Chunky Chicken, Ross Poultry in Aberdeen and sausage specialist Hall's of Broxburn. The expansion was financed by borrowings from the Bank of Scotland. The company later moved into beef and lamb and into territories as diverse as Northern Ireland, Thailand and the Netherlands.

According to Just-Food (Just-Food, 18-April-2005), Grampian aimed to consolidate its operations at Broxburn, i.e., the abattoir and factory purchased from Hall's, and create a centre of excellence for both pork processing and sausage production there, enabling the business to maintain and grow its position in an increasingly competitive marketplace. The pork processing operations at Buckie and part of the sausage manufacturing operations at Elmswell (Suffolk) were going to be re-located to Broxburn where increased throughput and production was going to create 200 new full-time jobs.

Despite what seemed to be an important expansion, the Broxburn abattoir did not survive, being sold to the Dutch conglomerate Vion for £380 million. The systemic instability, however, continued as in October 2012, Vion announced that Hall's of Broxburn would shut down, with a loss of 1700 jobs. Months of negotiations between the Dutch company and a Scottish Government-led taskforce ended in failure. Whilst the abattoir was demolished for redevelopment, Browns Food Group, based in Dumfries and Galloway, bought Hall's brand and recipes.

Despite the relatively short-lived ownership of Broxburn abattoir by Vion, it was during its tenure that an example of vertical collaboration (i.e., The PorkLink chain) was created between the Scottish Pig Producers Ltd., Vion Food Scotland Ltd. (Halls) and the UK multiple retailer ASDA. The chain won a Scotland Food and Drink award in the "Success through working in partnership" category in 2010. Moreover, it was the origin of the brand label "Specially Selected Pork" (Leat & Revoredo-Giha, 2013b).

The above episode was interesting because the profitability of the pig meat supply chain has been under constant pressure due to short-term variation in prices, and longer-term price depression, as well as escalating feed prices. In addition, the profitability was affected by issues related to production shocks in the form of animal diseases and to changes in animal welfare policy. A detailed analysis of the vertical collaboration on resilience and value creation can be found in Leat and Revoredo-Giha (2013a, b).

2.5.2.3 Brechin Abattoir

Since the closure of Vion's Broxburn plant, pigs had to be sent to England's abattoirs for processing and many weaner pigs had to be sold to English farmers for finishing. It was then that there was the hope that the Brechin abattoir could reduce transport distances, enhancing animal welfare standards and improving the logistics for the Scottish pig producers.

The abattoir at the Brechin site (Angus) was built in 1982 by the company AP Jess. In 2011, meat processor Tulip Foods (a Danish company) agreed on a deal with AP Jess to slaughter pigs at Brechin. In 2014, the abattoir was purchased by Quality Pork Ltd. (QPL). QPL was formed through close collaboration between the two pig cooperatives: Scottish Pig Producers and Scotlean, and the processor Tulip Ltd., to develop a modern processing facility for the Scottish pig industry.

After a refurbishing of £10 million, the Brechin abattoir opened in 2016 employing 58 people and providing much needed additional capacity to the pig meat industry in the country. The site now boasts a lairage, kill line, rapid carcass chill system and lorry wash. It has the capacity to process all Scottish-born pigs, with a capacity of more than 7500 pigs and 300 sows a week. In contrast with the Broxburn plant, which was an abattoir and processor, Brechin only deals with slaughtering and primal cuts. Pork was then delivered to Tulip Ltd's network of sites to be further processed into a wide variety of quality food products.

In August 2020, Tulip Ltd. announced its intentions to rebrand as Pilgrim's Pride Limited. This was due to the fact that Pilgrim's Pride acquired Tulip in August 2019 and the rebrand was part of Tulip's integration plan.

Whilst Britain exiting the European Union did not affect the abattoir operation, the COVID-19 situation did. Moreover, in 2021, due to an outbreak of COVID, the company gave up a lucrative export licence with China, which has not been reinstated by China, despite the plant getting the all-clear by public health authorities a few weeks after the outbreak. According to Pilgrim's, the situation of the QPL abattoir in Brechin became unsustainable.

In February 2022, the QPL abattoir was acquired by food manufacturer Browns Food Group (who also acquired Hall's brand from Vion in 2012). The same year, Browns bought John Robertson & Sons Hamcurers and Robertsons Fine Foods, ensuring that Browns is currently the largest pork processor in Scotland.

2.6 Discussion

The purpose of this section is to discuss the role that collaboration has played on the resilience and value creation of the pig supply chain. This is represented in Fig. 2.6.

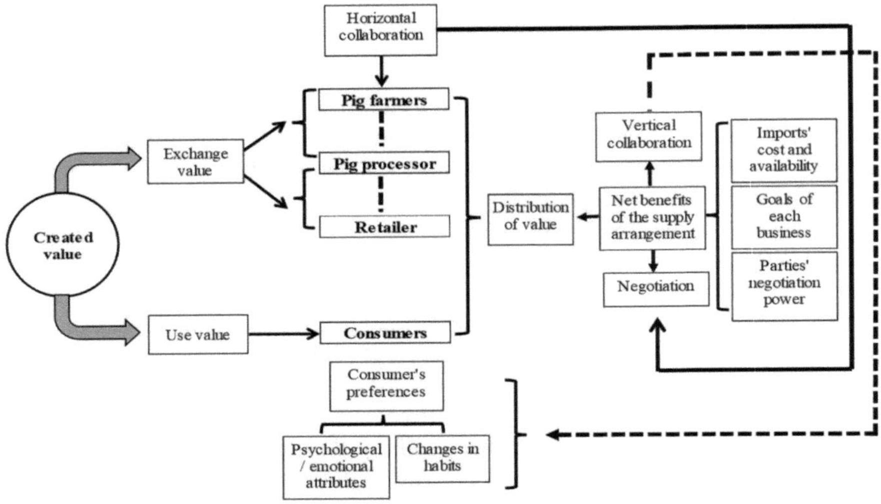

Fig. 2.6 Collaboration, resilience and value creation in the Scottish pig supply chain.

2.6.1 *Collaboration and Resilience*

Pig production in Scotland and elsewhere is challenging as producers are exposed to many significant disruptions. According to Quality Meat Scotland, pork producers were affected by the Russia-Ukraine war, which led the increase of feed cost by 70% compared to production cost in late 2021. This situation eroded market profits and rendered some businesses unsustainable.

Assessing the relevance of collaboration in the face of uncertainty due to the Covid-19 pandemic, Lotfi and Larmour (2022) concluded that the more firms collaborate (either horizontally or vertically) in the supply chain, the more resilient they are.

The Scottish Association of Meat Wholesalers (SAMW) emphasize that collaboration in the meat sector is important for building resilience in the face of disruptions (Meat Management, 2022; Medel et al., 2020).

In the midst of the supply chain disruptions described in the previous sections, it is clear that horizon collaboration has been important for the pig producers in Scotland. As shown in Fig. 2.6, it has allowed to create a supply of pigs of enough size and quality that would allow them to market not only in Scotland but also in England due to the fact that processors require enough raw materials (i.e., in this case pigs) of the right quality, complying with animal health and welfare standards.

As pointed out in Leat and Revoredo-Giha (2013a), in the case of the Scottish pig cooperatives, the horizontal collaboration between producers has enabled them to share views on how to innovate and take production forward, exchange production performance information, and provide the scale and reliability of supply that is important to large customers. It has also been developed between the producers and their representative organisations (i.e., between the pig cooperatives and Quality Meat Scotland).

At the producer level and beyond, there has been a readiness to anticipate and recognise the risks to be faced and to manage them accordingly. This risk management focus arises from the difficult experiences of producers, their cooperatives and pig processors in recent years (e.g., disease problems, market collapses, processing plant closures). The very proactive nature of the pig cooperatives' management has played an important role in developing a forward-looking approach of management (e.g., in planning waste management and examining the issue of energy efficiency in pig housing). This is quite important because it allows the cooperative to better negotiate with processors as they are a larger entity and they have other outputs to offer (i.e., not just the pigs).

Vertical collaboration involving the entire supply chain, although very important, has been very elusive. In the only episode of vertical collaboration along the entire supply chain (i.e., producers, processor and retailer), Leat and Revoredo-Giha (2013a, b) showed that it improved the producers' resilience by helping them reduce market risk.

It should be noted that the UK Net Zero strategy, which aims at ending the UK's domestic contribution to man-made climate change by 2050, provides an additional incentive for the establishment of collaboration along the supply chain (i.e., vertical collaboration) (Revoredo-Giha, 2024). This can be seen in the so-called 'Scopes', which are the basis for mandatory Green House Gas (GHG) reporting in the UK. There are three scopes: Scope 1 emissions—This covers the GHG emissions that a company makes directly, for example while running its boilers and vehicles; Scope 2 emissions—These are the emissions a company makes indirectly, such as when the electricity or energy it buys for heating and cooling buildings is being produced on its behalf; and Scope 3 emissions—These are all the emissions associated not with the company itself, but the organisation that it is indirectly responsible for, up and down its value chain. For example, by buying products from its suppliers (e.g., processors and farmers). Scope 3 nearly always involves the biggest GHG emissions.

The difficulty in establishing vertical collaborations for farmers is to convince the downstream potential partners (e.g., processors and retailers) of the advantages of the relationships whilst being willing to ensure farmers of the stability in prices and demand. This is due to the possibility for downstream members to procure supply from other parts (e.g., imports), which gives them greater negotiation power. Moreover, it is more difficult to establish specific arrangements when the suppliers operate in a variety of supply chains. However, despite the aforementioned difficulties, the requirement to reduce the Scope 3 GHGs makes processors and retailers (or any other potential partners downstream the supply chain, like wholesalers) see farmers in a different light, as they need farmers to reduce their GHGs, too.

The acquisition of Brechin by Browns Food Group improves the resilience of the Scottish pig supply chain and the resilience of the farmers' businesses. As noted in the historical section, after the closure of Lawson's of Dyce, the life expectancy of the processors' business has been short, implying a systemic risk particularly affecting farmers

2.6.2 Collaboration and Value Creation

An interesting aspect of the formulation of value in terms of value in exchange and value of use (see Fig. 2.6) is that it allows to accommodate several factors present in the decisions of stakeholders in the supply chain, such as the benefits of the supply system (e.g. price, information provision, efficiency in the logistics process, reliability of supply), the risks involved and the opportunity costs of the supply chain arrangement.

As pointed out in Leat and Revoredo-Giha (2013b), value within the supply chain (i.e. value in exchange) was created through farmers engaging in horizontal and vertical collaboration. For farmers, this value arises from efficient pig collection and transport, improved market and price security, and information on improving pig performance. For the processor and retailer, value is generated in terms of the improved security of the supply of an assured quality and the ability to communicate with their pig meat supply base. For consumers, value (i.e. value in use) comes through market development associated with a label that communicates clear provenance, high animal welfare practices and consistent product quality.

It is important to recognise that some of the value arising from the horizontal collaboration is a supply arrangement where the producer cooperatives have a set of goals which are related to those of their producer members. As marketing cooperatives, they seek to arrange the efficient delivery of its members' pigs to profitable and secure markets, which provide prompt regular payments. It is also seeking to aid its members in their production activities wherever possible, through providing market feedback and assisting knowledge transfer on pig production innovation.

In the past, when vertical collaboration was also in place, the retailer (i.e., ASDA) regularly attended producer meetings and shared information with the pig cooperatives on market development and other matters. The processor (i.e., Vion) also collaborated with the retailer and the pig producers in new product development so as to get a more balanced use of the pig carcass (i.e. through making use of the less popular parts of the carcass). In fact, to help meet their pricing obligations under PorkLink, Asda and VION worked together on the development of new products, helping to ensure a balanced use of the carcass. This means a reduction in waste and ensures that the whole animal is sold into a premium market, the validity of which were assisted by the Specially Selected Pork label and Scottish SPCA approval.

It should be highlighted that the purchase by Browns Food Group of the Brechin abattoir where the two cooperatives deliver pigs and the fact that Browns Food Group owns the Hall's brand, creates the possibility of creating a closer alliance between the cooperatives and the processor; as mentioned, Scope 3 may provide incentives to further negotiate with retailers. In this sense, it is important that Brown's Food Group is interested in increasing the availability of Scottish pork in major retailers (The Grocer, 2024). Although it is not possible to revert the trends in terms of the consumption of red meat, this has the advantage that it may stabilise the demand that pig producers face, as well as help with the competition from imported pork products. Moreover, Browns Food Group is also looking to increase production of Scottish Quality Meat Scotland 'Specially Selected Pork' for the retail sector in Scotland, helped by the fact it now had a fully integrated processing supply chain, from primary through to added-value.

The other additional aspect of value creation (in terms of exchange value) is that the acquisition of the Brechin plant complements the company's existing network of six processing sites within Scotland's central belt and streamlines the logistics of the supply chain by helping to cut half a million road miles of transport a year (The Grocer, 2024). As mentioned, prior to the acquisition, Pilgrim's UK would slaughter Scottish pork at Brechin, before transporting primal cuts to England to be further processed.

2.7 Final Remarks

Overall, a number of long-term factors have affected the performance of the Scottish pig supply chain, such as changes in consumer preferences (e.g., breakfast, health-oriented) which have decreased the per capita demand for pork and bacon over time, which have affected the final demand for the major pig products.

Another factor has been the increases in retailing concentration, which have changed the power balance along the supply chain. The possibility of importing pig products by retailers has also affected their power within the supply chain and the distribution of value. Whilst retailers were relatively small or local grocers, Scottish pig processors invested in expanding their businesses and increased the resilience of the Scottish supply chain. This situation changed when retailers increased their negotiation power to the detriment of the processors' power.

On the one hand, the closure of processing facilities in Scotland triggered horizontal collaboration at the farm level, which has been successful in keeping farmers afloat despite the disruptions on the supply chain. Vertical collaboration, on the other hand, has been more elusive and worked only for a short period (2009–2012) due to the processing sector pulling out from the arrangement. This may indicate that when the created values are reduced or less fairly distributed, the parties may have less incentive to maintain the vertical collaboration arrangements.

The search for vertical collaboration arrangements is important because it has the potential to improve the producers' resilience to shocks by taking into account changes in the costs of production, ensuring a stable demand, and to develop new products that will reinforce the collaboration.

Two current developments have the possibility to foster vertical collaboration within the Scottish pig supply chain. One is the purchase of Browns Food Group of the Brechin abattoir, the largest in Scotland. The processor is not only interested in expanding the use of Scottish pork in its operations, but also to use the label 'Specially Selected Pork', which could link the farmers' raw material with the final user (i.e., consumers).

The other possibility comes from the requirements behind the UK net zero strategy by 2050. This is, in particular, those related to Scope 3, i.e., emissions from sources connected to a business, rather than being generated by a business directly. The need to reduce those emissions may provide another opportunity to generate vertical collaboration for the Scottish pig supply chain and to include not only processors but also retailers in it.

Acknowledgments This chapter is part of the Scottish Government 2022–27 Strategic Research Programme (SRP) and comprises several projects dealing with food supply chains in Topic B4 (Food Supply and security) and Topic B5 (Food and drink improvements). The authors would like to thank Andy McGowan (Scottish Pig Producers) and Alan Stevenson (SAOS) for the comments and information at different stages of the project.

References

Barratt, M. (2004). Understanding the meaning of collaboration in the supply chain. *Supply Chain Management: An International Journal, 9*(1), 30–42.

Bodie, Z., & Merton, R. C. (2000). *Finance*. Prentice Hall.

Borys, B., & Jemison, D. B. (1989). Hybrid arrangements as strategic alliances: Theoretical issues in organizational combinations. *Academy of Management Review, 14*(2), 234–249.

Brandenburger, A. M., & Stuart, H. W., Jr. (1996). Value-based business strategy. *Journal of Economics and Management Strategy, 5*(1), 5–24.

Briano, E., Caballini, C., & Revetria, R. (2009, October). Literature review about supply chain vulnerability and resiliency. In *Proceedings of the 8th WSEAS international conference on system science and simulation in engineering* (pp. 191–197). World Scientific and Engineering Academy and Society (WSEAS).

Buckie Heritage. (Various years). https://buckieheritage.org/

Burt, S. L., & Sparks, L. (2003). Power and competition in the UK retail grocery market. *British Journal of Management, 14*(3), 237–254.

Christopher, M., & Peck, H. (2004). Building the resilient supply chain. *The International Journal of Logistics Management, 15*(2), 1–13.

Clarke, R. (Ed.). (2002). *Buyer power and competition in European food retailing* (Vol. 10). Edward Elgar Publishing.

Dania, W. A. P., Xing, K., & Amer, Y. (2016). Collaboration and sustainable agri-food suply chain: A literature review. In *MATEC web of conferences* (Vol. 58, p. 02004). EDP Sciences.

Davies, B. J., & Ward, P. (2000). Space allocation in UK grocery retailing. *British Food Journal, 102*(5/6), 406–419.

Dobson, P., Waterson, M., & Chu, A. (1998). *The welfare consequences of the exercise of buyer power*. Office of Fair Trading.

Edinburgh Evening News. (1940). Edinburgh Master Butchers' Association dinner – Thursday 22 February.

Fiksel, J. (2003). Designing resilient, sustainable systems. *Environmental Science & Technology, 37*(23), 5330–5339.

Helfat, C. E., Finkelstein, S., Mitchell, W., Peteraf, M., Singh, H., Teece, D., & Winter, S. G. (2009). *Dynamic capabilities: Understanding strategic change in organizations*. Wiley.

Hollingsworth, A. (2004). Increasing retail concentration: Evidence from the UK food retail sector. *British Food Journal, 106*(8), 629–638.

Insch, A. (2008). Triggers and processes of value creation in Australia's chicken meat industry. *British Food Journal, 110*(1), 26–41.

Just-Food. (2005). *K: Grampian invests at Broxburn, closes Buckie*. https://www.just-food.com/news/uk-grampian-invests-at-broxburn-closes-buckie/?cf-view&cf-closed

Lancaster, K. J. (1966). A new approach to consumer theory. *Journal of Political Economy, 74*(2), 132–157.

Leat, P., & Revoredo-Giha, C. (2013a). Risk and resilience in agri-food supply chains: The case of the ASDA PorkLink supply chain in Scotland. *Supply Chain Management: An International Journal, 18*(2), 219–231.
Leat, P. M. K., & Revoredo-Giha, C. (2013b). In search of differentiation and the creation of value: The quest of the Scottish pig supply chain. *British Food Journal, 115*(10), 1487–1504.
Lotfi, M., & Larmour, A. (2022). Supply chain resilience in the face of uncertainty: How horizontal and vertical collaboration can help? *Continuity & Resilience Review, 4*(1), 37–53.
Manuele, F. A. (2005). Risk assessment & hierarchies of control. *Professional Safety, 50*(5), 33–39.
Marks, H. F., & Britton, D. K. (Eds.) (1989). *A hundred years of British food & farming. A statistical survey* (p. xiv+, 275pp), Taylor and Francis Ltd.
Mason-Jones, R., & Towill, D. R. (1998). Shrinking the supply chain uncertainty circle. *IOM Control, 24*(7), 17–22.
McIntyre, T., Wilson, M., Schulze-Ehlers, B., & Childerhouse, P. (2022). *The governance of value creation and capture in agri-food supply chains: A conceptual framework, propositions and protocols*. https://agresearch.figshare.com/articles/report/The_Governance_of_Value_Creation_and_Capture_in_Agri-Food_Supply_Chains_A_Conceptual_Framework_Propositions_and_Protocols/26002018?file=46945165
Meat Management. (2022, June). *SAMW calls for meat supply chain collaboration*. Meat Management.Com.
Medel, K., Kousar, R., & Masood, T. (2020). A collaboration–resilience framework for disaster management supply networks: A case study of The Philippines. *Journal of Humanitarian Logistics and Supply Chain Management, 10*(4), 509–553.
Nicolaas Bezuidenhout, C., Bodhanya, S., & Brenchley, L. (2012). An analysis of collaboration in a sugarcane production and processing supply chain. *British Food Journal, 114*(6), 880–895.
Pettit, T. J. (2008). *Supply chain resilience: Development of a conceptual framework, an assessment tool and an implementation process* (Doctoral dissertation). The Ohio State University.
Ponomarov, S. Y., & Holcomb, M. C. (2009). Understanding the concept of supply chain resilience. *The International Journal of Logistics Management, 20*(1), 124–143.
Quality Meat Scotland. (2023). *The Scottish red meat industry profile*. Quality Meat Scotland.
Revoredo-Giha, C. (2024, June 10). Carbon neutrality requirements may bring vertical collaboration to the pig supply chain. *SEFARI Blog*. https://sefari.scot/blog/2024/06/10/carbon-neutrality-requirements-may-bring-vertical-collaboration-to-the-pig-supply
Smith, A. (1999). *The wealth of nations, books I-III*, edited by Andrew Skinner. Taylor & Francis Ltd
Scottish Government. (2021). *Pig sector – Reducing greenhouse gas emissions: Bringing home the bacon*. https://www.gov.scot/publications/bringing-home-bacon-reducing-greenhouse-gas-emissions-scottish-pig-sector/pages/6/
Steele, B., & Feyerherm, A. (2014). Loblaw sustainable seafood: Transforming the seafood supply chain through network development and collaboration. In *Building networks and partnerships* (pp. 101–132). Emerald Group Publishing Limited.
The Grocer. (2024, May 20). *Browns food group targets £100m sales jump for Scottish pork*.
The Press and Journal. Aberdeen, Aberdeenshire & Highlands news (issues 10-7-1965 and 24-02-1976
The Scotsman. (2021). *Scotsman obituaries: David R Hall OBE, man who built Hall's butchery brand*. https://www.scotsman.com/news/people/scotsman-obituaries-david-r-hall-obe-man-who-built-halls-butchery-brand-3231998
The Scottish Farmer. (2022, March 17). *Pig sector in crisis*. https://www.thescottishfarmer.co.uk/news/19999877.pig-sector-crisis/
Wileman, A., & Jary, M. (1997). *Retail power plays: From trading to brand leadership* (p. 280). New York University Press.
Wrigley, N. (1987). The concentration of capital in UK grocery retailing. *Environment and Planning A, 19*(10), 1283–1288.
Wrigley, N. (1991). Is the 'golden age'of British grocery retailing at a watershed? *Environment and Planning A, 23*(11), 1537–1544.

Chapter 3
Consumers' Attitudes Towards Visiting a Wine Destination

Spyridon Mamalis, Irene Kamenidou, Elisavet Bouloubasi, and Aikaterini Karampatea

3.1 Introduction

Wine tourism has gained significant attention as a form of alternative tourism in recent years; especially since the mid 90's (Carrà et al , 2016), it is considered as one of the most important and promising types of tourism (Figueroa & Rotarou, 2018). However, there is no united definition and conceptualization of the term 'wine tourism'. For example, Hall et al. (2002, p. 3) defined wine tourism as "visitation to vineyards, wineries, wine festivals and wine shows for which grape wine tasting and/or experiencing the attributes of a grape wine region are the prime motivating factors for visitors". Brown and Getz (2005, p. 266), state that wine tourism involves traveling specifically to visit wine-producing regions or visiting these areas while traveling for other reasons, and this type of tourism is popular among "wine lovers" (Connolly, 2019). According to the winelab.eu (2018), wine tourism is "A service experience; A complex experience, involving the whole *terroir*'s attractions; Related to the wine tourist's different motivations and profiles; Nowadays managed through electronic marketing tools; A tool for regional development (e.g. wine routes)."

The convergence of wine production and tourism has created a dynamic industry that attracts millions of tourists annually. Wine tourism has become a vital component of rural and regional tourism as its significance lies in its ability to generate revenue (Festa et al., 2020). It also supports local economies directly and indirectly. Wine tourism contributes significantly to the economic growth of regions and nations (Kastenholz et al., 2023; Hall et al., 2004), by creating jobs, supporting small businesses, and stimulating ancillary industries such as hospitality,

S. Mamalis (✉) · I. Kamenidou · E. Bouloubasi · A. Karampatea
Department of Viticulture and Oenology, Democritus University of Thrace, Drama, Greece
e-mail: mamalis@vo.duth.gr; eikamenidou@vo.duth.gr; bouloumpasi@gmail.com; katerina_karampatea@yahoo.gr

K. Mattas et al. (eds.), *Biodiversity in Agri-Food Systems*, Cooperative Management, https://doi.org/10.1007/978-3-032-06385-4_3

transportation, and retail (Brito et al., 2024; Santos et al., 2022). Moreover, wine tourism promotes cultural preservation (Kastenholz et al., 2023) because many wine-producing regions are often steeped in rich histories and traditions that are intrinsically linked to viticulture. As wine destinations, they actively promote local heritage among visitors, showcasing both historical and contemporary advancements in winemaking practices. Additionally, wine tourism promotes cultural exchange (Vîlcea et al., 2024), through interaction with international tourists, bringing diverse perspectives and experiences from tourists; on the other hand, tourists gain insights into the local way of life. Likewise, there is an educational aspect of wine tourism that holds significant value. Tours of vineyards and wineries typically include information on winemaking techniques, historical practices, and the cultural significance of wine in the region (e.g., Hernández et al., 2022). This enhances the visitor's experience while also transferring knowledge to younger generations. Furthermore, since the wine tourism industry aligns with the desired sustainable practices globally, it contributes to a sustainable future either by organic production or by decreasing the ecological, water, carbon, and energy footprint (e.g., Trigo & Silva, 2022; Bole, 2024).

Regarding Greece, wine tourism and production are vital to Greece's culture and economy. With over 4000 years of winemaking history (Estreicher, 2017), wine production and wine tourism are key to Greece's economic growth and promoting rural development (e.g., Karampatea et al., 2024; Pitoska et al., 2023; Alebaki & Iakovidou, 2010a).

In order for wine tourism to be successful in the Greek realm, Greek wine tourist destinations need to acquire knowledge of the tourists' motives to engage in wine tourism, as well as the critical factors for choosing a specific wine destination rather than others. Researchers have stated that wine tourists' behavior differs from one individual to another, while scholars have recognized the absence of a uniform or stereotypical classification of wine tourists (Charters & Ali-Knight, 2002). This highlights the importance of employing a segmentation-based approach to developing effective marketing strategies. Thus, this study aims to explore the factors that tourists prioritize when deciding to visit a wine destination, as well as their motives, and seeks to segment visitors based on these elements. To accomplish this goal, this research utilizes data collected from Greek wine tourists.

3.2 Literature Review

There is an abundance of studies that have examined wine tourism from various perspectives, such as at the macro and micro levels (e.g., Sun & Drakeman, 2022; Alant & Bruwer, 2010), from the viewpoints of companies and stakeholders (e.g., Festa et al., 2020), and from the tourists' perspective (e.g., Ferguson et al., 2018). There are plentiful studies on wine tourism and consumer segmentation based on different variables, such as expenditures (Saayman et al., 2012; Marujo et al., 2022), constraints (Gu & Huang, 2019; Cho et al., 2017; Bonn & Cho, 2013), lifestyle

(Hernández et al., 2022; Simpson et al., 2005; Simpson & Bretherton, 2004), generational cohorts (Stergiou et al., 2018; Kruger & Viljoen, 2022; Seyedimany & Koksal, 2022), demographic characteristics (Alebaki & Iakovidou, 2011; Seyedimany & Koksal, 2022), and involvement (Santos et al., 2020; Koksal, 2021; Gu et al., 2018; Nella & Christou, 2014, 2021), just to name a few.

As to the focus of this study, which is segmentation based on motives to visit a wine destination, and critical factors for choosing a specific wine destination, the latter seems to be understudied. Segmentation based on motives for wine tourism refers to either a region (Hernández et al., 2022) or winery (Nella & Christou, 2021) and there are quite a large number of studies that deal with this matter (e.g., Vorobiova et al., 2020; Kelley et al., 2019; Payini et al., 2022; Napolitano et al., 2022). As regards critical choice factors for a specific wine tourism destination (either region or winery), a small number of studies exist (Mamalis et al., 2023; Cohen & Ben-Nun, 2009; Masberg, 2014).

Kim et al. (2019) studied tourists' wine-related motivations (N = 267), constraints associated with wine travel, and perceptions of wine tourism destination attributes (n = 10 wineries in the Italian Sardinia Region). Upon these variables, they employed a motivation-based market segmentation approach; through cluster analysis, three segments were detected: the "Wine lovers" (n = 138), the "Wine culture tourists" (n = 81), and the "Casual wine tourists" (n = 48). Alebaki and Iakovidou (2010b) profiled and segmented visitors (N = 133) to Greek wineries (N = 13) along the "Wine Roads of Northern Greece". They found that the typical winery visitor in Northern Greece is predominantly a young, well-educated male with a high income, residing in urban areas near the wine region. Through a Two-step Cluster Analysis based on the wine tourists' motivations for visiting the region, they identified four distinct groups: "Wine Lovers," "Neophytes," "Occasional Visitors," and "Hangers-on". Bruwer et al. (2018) examined the motivations of wine tourists to visit winery cellar doors and their perceptions of destination image. The study, which involved 676 visitors across 17 winery cellar doors (Barossa Valley Wine Region of South Australia), identified five distinct motivational clusters: "Wine learners" (7%), "Dining enthusiasts" (21%), "Wine buyers" (49%), "Wine enthusiasts" (4%), and "Wine connoisseurs" (19%). Napolitano et al. (2022) segmented wine tourists (N = 394; South Australia) based on their perceptions of their social interactions and found two segments: the "Social interaction enthusiasts" and "Travel companion-oriented interactors". Vorobiova et al. (2020) investigated tourists' motives (N = 217) visiting La Rioja, Spain and segmented visitors based on five derived motivational factors, namely: "Local attractions", "Wine-related motivations", "Adventure and sports", "Price and distance", and "Nightlife and shopping". Segmentation analysis identified two distinct clusters, the "Wine tourists" and the "Other tourists". Choi et al. (2017) explored the motives, wine purchasing involvement, and purchasing behavior of winery tourists (N = 302; Niagara region, Canada) and proceeded to market segmentation. Exploratory factor analysis for motives extracted 4 factors: "Escape and relaxation"; "Wine knowledge and education"; "Enjoying wine and the winery"; and "Socializing and kinship". Exploratory factor analysis for involvement extracted 3 factors: "Brand decision", "Purchasing", and “Product”. Segmentation produced three distinct groups: the “Wine Enthusiast” (n = 192), the “Wine Novice” (n = 57), and the “Wine Interested” (n = 54).

Masberg (2014) identified five key attributes for wine tourism in the Red Mountain American Viticultural Area: “Knowledgeable winery staff”, “Desirable lodging options”, “Clear directional signage”, “Wineries open daily with regular hours”, and “Open year-round”. Additional factors which were deemed significant included: “Personalized and unique experience (e.g., uncrowded)”, “Quality of the product and the perception that the cost matched the price (e.g., specialty wine, free or limited wine tasting, fees)”, “Having additional amenities which create a complete experience (e.g., food and wine pairings, local and unique charm, scenic/rural atmosphere, chance to talk to knowledgeable winery person, adequate parking)”, and “Choice of wines as well as experiences”. Cohen and Ben-Nun (2009) classified 373 visitors to Israeli wineries by 42 key features of wineries and wine regions that critically influence visit decisions, as assessed using a 5-point Likert-type scale. Through factor analysis, three key factors emerged: “Winery atmosphere,” “Cultural activities,” and “Family activities”. The subsequent segmentation analysis identified seven distinct visitor segments: “Subjects who have not visited a winery during the last two years”; “Subjects who have visited a winery during the last two years”; “Respondents who have children”; “Respondents who do not have children”; “Low wine-involved consumers”; “Medium wine-involved consumers”; and “High wine-involved consumers”. Mamalis et al. (2023) investigated the key factors of a winery and its surrounding area (n = 47) that are important to wine tourists (n = 430) when selecting a winery to visit. The analysis initially identified the primary elements: “The winery offers a tour of the production area, cellars, etc.,” “The winery offers wines for retail sale,” and “The winery offers a vineyard tour”. Factor analysis of the 47 elements yielded four factors, namely “Winery terroir elements”, “Core elements of winery tourism”, “Experiential learning” and “Educational activities”. The segmentation analysis based on the four dimensions revealed three distinct segments: the “Demanding Winery Visitors,” the “Core-Centered Winery Visitors,” and the “Do Not Care Winery Visitors”.

3.3 Methodology

The research employed an online data collection method using a structured questionnaire distributed through digital platforms. This methodology was chosen for its efficiency in reaching a broad and diverse audience, cost-effectiveness, and ability to collect data quickly (Evans & Mathur, 2005). Items and questions were adopted from previous research (e.g., Hernández et al., 2022; Kim et al., 2019; Brandano et al., 2019; Bruwer et al., 2018; Choi et al., 2017; Alebaki et al., 2015; Alant & Bruwer, 2004). The questionnaire was designed using an online survey platform (Google Forms) comprising of closed-ended questions, including Likert scale and Likert-type scale items, multiple-choice questions, and demographic questions. The questionnaire was pre-tested with a small group of respondents to identify and rectify potential issues related to clarity and usability. The non-probability (convenience, criteria) sampling method was utilized for participant recruitment. Participants were provided with a clear introduction explaining the study's purpose, ensuring informed consent, voluntary participation and anonymity. Additionally, the first question was related to consent for using the participants' data. Data was collected during a six-month period, from March to August 2024, and was exported to statistical software for analysis (descriptive statistics, factor analysis, reliability of factors and total scale, hierarchical and k-means cluster analysis, and chi square tests).

3.4 Results

3.4.1 *Sample Profile*

Females were overrepresented, while the age groups 18–29, 30–39 and 50–59 had quite similar percentages (24.4%, 25.2% and 26.7% respectively). In general, the sample could also be considered highly educated, since 69.9% had at least a university degree. Additionally, half of the sample was married (48.1%) and had an annual individual net income ranging between 10,000 and 19,999 euros (Table 3.1).

Table 3.1 Socioeconomic data of participants

Socioeconomic data of participants		N	%
Gender	Male	120	45.1
	Female	146	54.9
Age	18–29	65	24.4
	30–39	67	25.2
	40–49	53	19.9
	50–59	71	26.7
	60+	10	3.8
Marital status	Single	72	27.1
	In a relationship	50	18.8
	Married	128	48.1
	Divorced/widowed	16	6.0
Education	Secondary education—High school	43	16.2
	Postsecondary education	37	13.9
	Bachelor degree	133	50.0
	Master/PhD	53	19.9
Net individual income per year in euros	Up to 9999.99€	72	27.1
	10000.00–19999.99€	134	50.4
	20000.00–29999.99 €	47	17.7
	30000.00 + €	13	4.9

3.4.2 Interest in Wine Tourism, Critical Visitation Factors in Wine Tourism Destination Choice, Motives for Wine Tourism, Satisfaction and Recommendation

The results indicate that tourists generally display a moderate interest in wine tourism, with a mean value (MV) of 2.98 (Std = 0.83). Specifically, 4.1% and 18.8% reported having no interest or low interest in wine tourism, respectively, while the majority (56.8%) expressed a moderate level of interest. Moreover, 15.8% showed a high and 4.5% a very high interest in wine tourism.

Table 3.2 outlines, in percentages (%) and mean values (MV) with its standard deviation (Std), the critical factors influencing the choice of a wine destination (presented in descending order of importance). The three most significant factors identified are: the availability of sufficient information about the destination and its wineries (MV = 4.09), the presence of visitable wineries (MV = 4.03), and the availability of wine-tasting experiences for visitors at the wineries (MV = 3.95). The numbers 1–5 in the first row of Table 3.2 represent the answers of the Likert scale.

The three primary motives for visiting a wine destination (Table 3.3) are gaining the ability to recognize quality wines (MV = 3.95), experiencing local wines and cuisine (MV = 3.91) and acquiring knowledge to make informed wine purchases (MV = 3.90). The numbers 1–5 in the first row of Table 3 represent the answers of the Likert scale.

Visitor satisfaction at the chosen wine tourism destination was reported as follows: 21.4% were neutral, 75.2% satisfied, and 3.4% very satisfied (MV = 3.82; std. = 0.47). Regarding the intention to recommend the wine tourism destination, 0.8% were very unlikely, 4.5% were unlikely, and 15.4% were neutral. Meanwhile, 54.4% were likely and 25.2% were very likely to recommend it (MV = 3.99; std. = 0.81).

3.4.3 Factor Analysis and Segmentation

The 15 choice factors and 14 motivational factors for visiting a wine destination were analyzed using Exploratory Factor Analysis (EFA) with Principal Component Analysis (PCA) and varimax rotation. Factors with eigenvalues greater than 1.00 (Kaiser, 1958) and items with their loadings >0.4 were retained (Ford et al., 1986). Conversely, items with cross-loadings on multiple factors or loadings below 0.4 were excluded. As a result, the factor analysis for critical choice factors led to the elimination of seven items, whereas no item was removed from the analysis of motivational factors for visiting a wine tourism destination.

Table 3.2 Critical factors in wine tourism destination choice

Critical choice factors	1	2	3	4	5	MV	Std
There should be enough information available about the destination and wineries	1.9	4.9	7.1	54.5	31.6	4.09	0.86
Wineries accept visitors	0.8	4.5	7.9	64.3	22.6	4.03	0.74
Wineries hold wine-tasting courses	1.5	6.0	12.0	56.4	24.1	3.95	0.86
Wineries have a parking space	2.3	9.0	12.4	55.6	20.7	3.83	0.93
Wineries have extended hours of operation	3.0	4.1	18.0	59.8	15.0	3.80	0.85
There are organized tourist accommodations in the area	2.6	8.3	17.7	56.4	15.0	3.73	0.91
The wineries have a unique atmosphere	1.5	5.6	21.1	63.2	8.6	3.72	0.76
The area that the winery is located in is characterized by its natural wealth	3.4	6.8	21.8	55.6	12.4	3.67	0.90
Wineries hold wine production courses	4.1	5.6	22.9	58.3	9.0	3.62	0.88
The region is popular for its wines	1.5	11.3	26.3	52.3	8.6	3.55	0.86
Organized wine tourism trips are carried out in the region	4.1	10.9	16.5	63.9	4.5	3.54	0.90
Wineries allow tourists to participate in the production process	7.5	7.5	28.9	46.2	9.8	3.43	1.02
The area should be located on well-known wine routes	6.0	9.4	33.8	44.7	6.0	3.35	0.95
Wineries allow tourists to participate in the harvest process	10.9	9.4	24.4	46.6	8.6	3.33	1.11
There are many wineries in the area to visit	9.8	10.2	35.3	43.2	1.5	3.17	0.98

Table 3.3 Motives for wine tourism

Motivates for wine tourism	1	2	3	4	5	MV	SD
To learn to recognize a good wine	1.9	7.1	6.8	62.4	21.8	3.95	0.86
To try local wines and associated local food	1.5	5.6	13.5	59.0	20.3	3.91	0.83
To learn how to purchase quality wines	3.4	5.6	9.0	61.7	20.3	3.90	0.90
To differentiate myself from the members of my circle	18.0	22.2	17.7	39.5	2.6	2.86	1.20
To get to know the local history and tradition of wine	1.5	5.3	18.4	58.3	16.5	3.83	0.82
To experience moments of relaxation and fun	2.3	4.9	17.3	58.6	16.9	3.83	0.85
To be informed about wine by competent people	2.6	5.3	13.5	65.4	13.2	3.81	0.83
To taste and buy wines	1.5	6.4	18.4	60.5	13.2	3.77	0.81
To learn how wine is made	1.5	7.1	25.9	54.5	10.9	3.66	0.82
To visit a wine route	3.8	7.9	22.9	52.3	13.2	3.63	0.94
To participate in a wine festival	2.6	12.4	16.2	57.1	11.7	3.63	0.94
To participate in activities that will fill my weekend or vacation	3.0	9.0	26.7	53.8	7.5	3.54	0.87
To connect with other wine lovers	4.9	14.3	30.8	42.5	7.5	3.33	0.98
To buy souvenirs and gifts	15.4	12.0	19.2	51.5	1.9	3.12	1.15

Table 3.4 Critical choice factors of a wine destination derived from factor analysis

	Factor	
Critical choice factors of a wine destination	1	2
The area should be located on well-known wine routes	0.892	
There are many wineries in the area to visit	0.859	
The area that the winery is located in is characterized by its natural wealth	0.750	
Organized wine tourism trips are carried out in the region	0.673	
The region is popular for its wines	0.667	
There are organized tourist accommodations in the area	0.640	
There should be enough information available about the destination and wineries		0.888
Wineries accept visitors		0.792

Regarding the wine destination critical choice factors, factor analysis using PCA with varimax rotation identified two factors (KMO = 0.845; Bartlett's Test of Sphericity = 1074.276, df = 28, $p < 0.001$) that accounted for 68.22% of the total variance (TV). The first factor, which explained 43.72% of the TV, had a mean factor score (MFS) of 3.50 (Std = 0.73) and was labeled "Wine Route Prominence and Amenities". The second factor, explaining 24.50% of the TV, had an MFS of 4.06 (Std = 0.71) and was labeled "Information Availability and Visitor Access" (Table 3.4). Table 3.4 presents the two factors, the item per factor, as well as the loading on the factor.

As regards motivation to engage in wine tourism, factor analysis with PCA varimax rotation (KMO = 0.824; BTS = 3345.964; df = 91; sign. <0.001) produced three factors explaining 74.58% of the total variance (TV). The first factor explains 33.73% of the TV, has MFS = 3.83 (std = 0.70), and is named "Wine knowledge". The second factor explains 21.57% of the TV, has MFS = 3.66 (std = 0.76), and is named "Experiential tourism". Lastly, the third factor explains 19.28% of the TV, has MFS = 3.11 (std = 0.96), and is named "Socialization" (Table 3.5). Table 3.5 presents the three factors and the item per factor, as well as the loading on the factor.

3.4.4 *Segmentation of Wine Tourists*

In the segmentation procedure, apart from the five factors extracted (two critical choice factors and three motive factors to engage in wine tourism) and specifically their MFS, two other variables were added, i.e., satisfaction from and intention to recommend the wine destination that participants visited. Hierarchical cluster analysis with the Wards method preceded K-means cluster analysis which offered a three-cluster solution (Hair et al., 2014). The three-segment solution is presented in Table 3.6, as well as its members per cluster (n), together with the Final Cluster Centers (FCC), and the results of the ANOVA tests (F-statistic and p-value). ANOVA tests revealed that in all cases, the three clusters differed in the dimensions analyzed.

Table 3.5 Motive factors of a wine destination derived from factor analysis

	Factors		
Items per factor	1	2	3
To taste and buy wines	0.654		
To try local wines and local meals	0.615		
To learn how wine is made	0.773		
To learn to recognize a good wine	0.840		
To learn to buy good wines	0.862		
To get to know the local history and tradition of wine	0.822		
To be informed about wine by competent people	0.862		
To experience moments of relaxation and fun		0.861	
To participate in a wine festival		0.842	
To visit a wine route		0.714	
To participate in activities that will fill my weekend or vacation		0.572	
To buy souvenirs and gifts			0.856
To differentiate myself from the members of my circle			0.828
To connect with other wine lovers			0.653

Table 3.6 Segments of wine tourists

	Clusters			ANOVA tests	
Dimensions/variables	1st n = 154	2nd n = 95	3rd n = 17	F	Sig.
Wine route prominence and amenities	3.88	3.21	1.71	134.570	<0.001
Information availability and visitor access	4.14	4.29	2.03	117.462	<0.001
Wine knowledge	4.13	3.69	1.92	159.398	<0.001
Experiential tourism	4.09	3.17	2.43	76.268	<0.001
Socialization	3.75	3.64	1.61	153.471	<0.001
Satisfaction	3.86	3.73	4.00	4.129	0.018
Intention to recommend	4.15	3.75	3.82	31.171	<0.001

Total sample = 266

Crosstabulation tests aided to explore the demographics of each segment. Based on the results of Table 3.5 and the crosstabs per segment, the segments were named as follows:

Segment No. 1: "Popular destination orientated tourist", comprising 57.9% of the winery visitors. It is the most committed to wine tourism, as it has almost all FCC higher than the other segments. It is also the most demanding one in terms of motivation to visit a specific wine tourism destination. Additionally, people in this group tend to be satisfied and will recommend the wine destination that they have visited. This segment consists mainly of males (82.4%), aged 40–49 (35.3%), married (64.7%), highly educated with having a bachelor's degree (47.1%), and with a net individual annual income up to 10000.00€ (47.1%). This segment is the one that has a higher intention to recommend the wine tourism destination compared to the other segments.

Segment No. 2: "Socializing orientated tourist", comprising 35.7% of the winery visitors, with the highest FCC on the second factor regarding critical choice that has to do with information availability of the wineries and if they accept visitors. As to motives, people in this group care about socializing and secondly about obtaining knowledge about wine and wine production. Additionally, they tend to be satisfied (MV > 3.51) and tend to be likely to recommend (MV > 3.51) the wine destination that they have visited, while having the lowest level of satisfaction and intention to recommend compared to the other segments. In this segment, females overrepresent males (55.8% and 44.2% respectively), with age ranging between 40–49 (43.2%), and are married (68.4%). They are the highest educated group with 48.4% having a bachelor's degree and 26.3% having a postgraduate degree or a PhD. Lastly, the members of this group have a net individual annual income ranging between 10000.00 and 19999.99€ (47.4%).

Segment No. 3: "Don't care but it was a good experience tourist", comprising 6.4% of the wine tourists with FCC ranging from 1.71–4.00. The element that has the highest FCC is experiential tourism, while they are the most satisfied amongst the segments. They don't care about wine tourism in general, since the FCC of the two critical factors for choosing a destination are low (FCC = 1.71 and FCC = 2.03); thus, they probably have low expectations, and their visit was better than expected, which resulted in satisfaction and intention to recommend. This segment's males and females are quite close in percentage terms (48.7% and 51.3% respectively). It consists of the younger age groups (18–29: 30.5% and 30–39: 37.0%), while single and married are almost equally represented (35.1% and 33.8%, respectively). Additionally, they are highly educated with 51.3% having a bachelor's degree and a net individual annual income ranging between 10000.00 and 19999.99€ (54.5%).

3.5 Discussion and Managerial Applications

This study aimed to examine the critical factors affecting wine tourism destination choice, motives for visiting, tourist satisfaction, and recommendation, as well as the potential segmentation of wine tourists. Using a sample of 266 Greek wine tourists, several insights were generated, contributing to the extant research on wine tourism behavior.

Regarding critical choice factors for engaging in wine tourism, 15 items were presented and rated. Findings revealed that the most critical factors for wine tourism are: "There should be enough information available about the destination and wineries", "Wineries accept visitors" and "Wineries hold wine tasting courses". Moreover, two main factors (from factor analysis) influencing the choice of a wine destination were extracted: "Wine Route Prominence and Amenities" and "Information Availability and Visitor Access". These elements correspond with earlier studies indicating that the accessibility of detailed information and well-structured facilities substantially impact tourists' choices of destinations (Masberg, 2014; Mamalis et al., 2023).

Considering motives for visiting a wine tourism destination, 14 motives were presented to participants, which they rated on an attitudinal scale. The highest rated motives for visiting a wine tourism destination were: "To learn to recognize a good wine", "To try local wines and associated local food" and "To learn how to purchase quality wines". Moreover, the factor analysis of these 14 items produces three motivational factors: "Wine Knowledge," "Experiential Tourism," and "Socialization". Acquiring wine knowledge, such as learning to recognize quality wines and understanding winemaking processes, resonates with findings from previous studies, such as Choi et al. (2017) and Bruwer et al. (2018). Additionally, the "socialization" factor highlights the role of wine tourism in promoting connections among wine enthusiasts, as evidenced by previous research (Napolitano et al., 2022; Choi et al., 2017; Vorobiova et al., 2020).

The segmentation analysis revealed three distinct groups of wine tourists. The first segment, the "Popular destination-orientated tourist", appears to be the most committed segment, prioritizing wine knowledge, information availability and visitor's access to amenities, being up to one point similar to the "Wine Interested" identified by Choi et al. (2017), and "Wine learners' identified by Bruwer et al. (2018). The second segment, the "Socializing-oriented tourists," appears to consist of moderately engaged wine tourists who prefer destinations that offer opportunities for socialization, with easy access and information, while enhancing their wine knowledge. This group could be akin to the "Social interaction enthusiasts" segment identified by Napolitano et al. (2022). The third segment, the "Don't care but it was a good experience tourist", values mostly the experiential side of wine tourism since they do not care for learning about wine, socializing with others, or the offered amenities while visiting the wine destination. This segment represents the least invested wine tourism group and can be compared (to some point) to the "Do Not Care Winery Visitors" in the study by Mamalis et al. (2023).

The findings of the present study also reveal moderate to high levels of satisfaction (MV = 3.82) and the likelihood of recommending the chosen wine destinations (MV = 3.99). Since satisfaction is important for positive WOM, recommendation, loyalty and likeliness of repeat visits, managerial implications should focus on increasing satisfaction of the critical choice factors as well as the key motives for wine tourism. Additionally, based on the segments derived as well as the main motives for wine tourism, and the critical factors in choosing a wine destination, managers should focus on communicating their wine destinations by connecting visitors with the local culture and history of a wine region. Showcasing traditional practices, unique winemaking processes, and regional cuisine aids in fostering a deeper understanding and appreciation for the destination and its products. Personalized wine tourism experience aligning with the preferences of various market segments, as well as individual tourists, is essential. Therefore, by analyzing wine tourists' profiles, wine destination managers can design tailored experiences that resonate with diverse interests and enhance overall satisfaction. Additionally, effectively exploiting digital marketing and social media is crucial, by creating and maintaining a successfull online presence, interacting with potential visitors, promoting their events, and sharing content that highlights the unique features of their offerings. Furthermore, wine destination marketing managers should invest in offering diverse and engaging activities, since previous research has revealed that wine tourism is connected to other activities apart from the core wine product itself. Therefore, offering activities beyond wine tasting, such as experiential learning (e.g., for young couples), with culinary workshops, cultural performances, or adventure-based activities, will motivate young potential tourists to engage in wine tourism and select a specific wine tourism destination. These are also essential for building a strong wine destination brand and creating loyalty programs and incentives.

3.6 Conclusion, Limitations and Directions for Future Research

The contribution of this empirical study to the prevailing research is that it highlights motives and key choice elements valued by Greek wine tourists, addressing a research gap in this area. Additionally, by identifying key factors, drivers, and segments, it offers a guide for improving the Greek wine tourism experience. This research faced limitations that suggest new directions in future research. Firstly, participation criteria like visiting a wine tourist destination and having internet access limit generalizations. Additionally, the study focused only on Greek wine tourists. Lastly, some potentially important variables for winery visitors were not considered, such as wine involvement.

Funding: This research was financed by the Winelab 2.

References

Alant, K., & Bruwer, J. (2004). Wine tourism behaviour in the context of a motivational framework for wine regions and cellar doors. *Journal of Wine Research, 15*(1), 27–37.

Alant, K., & Bruwer, J. (2010). Winery visitation sets: Intra-regional spatial movements of wine tourists in branded regions. *International Journal of Wine Business Research, 22*(2), 191–210.

Alebaki, M., & Iakovidou, O. (2010a). Initiatives towards wine tourism development in Greece. *Enometrica: Review of the Vineyard Data Quantification Society and the European Association of Wine Economists, 3*(2), 15–24.

Alebaki, M., & Iakovidou, O. (2010b). Segmenting the Greek wine tourism market using a motivational approach. *New Medit: Mediterranean Journal of Economics, Agriculture and Environment, 9*(4), 31.

Alebaki, M., & Iakovidou, O. (2011). Market segmentation in wine tourism: A comparison of approaches. *Tourismos, 6*(1), 123–140.

Alebaki, M., Menexes, G., & Koutsouris, A. (2015). Developing a multidimensional framework for wine tourist behavior: Evidence from Greece. *Wine Economics and Policy, 4*(2), 98–109.

Bole, D. K. (2024). Sustainable wine tourism: Best practices. In *Wine tourism and sustainability: The economic, social and environmental contribution of the wine industry* (pp. 95–122). Cham.

Bonn, M. A., & Cho, M. (2013). Constraints segmentation and wine tourism: Selecting target markets based upon consumers' attitudes and behavioral similarities related to their intent to visit wine destinations. https://d1wqtxts1xzle7.cloudfront.net/121036889/Bonn-Cho1-libre.pdf?1737944729=&response-contentdisposition=inline%3B+filename%3DConstraints_Segmentation_and_Wine_Touris.pdf&Expires=1759482407&Signature=XhIK99~6hO15f-E1jpK84g82pABvjfFRPXnD9LNG1AfOi-4rQSjpJo8mwkrF0~S-OKpqbAT3PDhO55Nd0rJuw1k6iVmdT0DuzqpuB~FLfj~qzbUW3nnLTXF6glzC2iO53BReqCl1knq3wsEt-piN8i7b-mDV3mBvvKj9et0ci0riEu-YwRPSKWqfPm1u25if7vIpfIRtP2t80b23-20a0WJf-WR0e4A~I9x8vqHaCY5wk8mqK~PgCRS8W3q2mlYHxef48-LqmKXFvE6N~ItZmo0IPril6h6G3Lr-LzmZ7HiSgOrMZJdhIYhCjrDqkt~aqRf4cCU-KJNsV2RgKzvd-dpzg__&Key-Pair-Id=APKAJLOHF5GGSLRBV4ZA

Brandano, M. G., Osti, L., & Pulina, M. (2019). How motivations and satisfaction influence wine tourists' loyalty? An analysis of the Italian case. *International Journal of Culture, Tourism and Hospitality Research, 13*(1), 55–69.

Brito, C., Pereira, S., Martins, S., Monteiro, A., Moutinho-Pereira, J. M., & Dinis, L. (2024). Strategies for achieving the sustainable development goals across the wine chain: A review. *Frontiers in Sustainable Food Systems, 8*, 1437872.

Brown, G., & Getz, D. (2005). Linking wine preferences to the choice of wine tourism destinations. *Journal of Travel Research, 43*(3), 266–276.

Bruwer, J., Prayag, G., & Disegna, M. (2018). Why wine tourists visit cellar doors: Segmenting motivation and destination image. *International Journal of Tourism Research, 20*(3), 355–366.

Carrà, G., Mariani, M., Radić, I., & Peri, I. (2016). Participatory strategy analysis: The case of wine tourism business. *Agriculture and Agricultural Science Procedia, 8*, 706–712.

Charters, S., & Ali-Knight, J. (2002). Who is the wine tourist? *Tourism Management, 23*(3), 311–319.

Cho, M., Bonn, M. A., & Brymer, R. A. (2017). A constraint-based approach to wine tourism market segmentation. *Journal of Hospitality and Tourism Research, 41*(4), 415–444.

Choi, C. H., Huang, S., Flaherty, J., & Khazaei, A. (2017). Segmenting wine tourists in Niagara, Ontario using motivation and involvement. *International Journal of Tourism Sciences, 17*(3), 198–212.

Cohen, E., & Ben-Nun, L. (2009). The important dimensions of wine tourism experience from potential visitors' perception. *Tourism and Hospitality Research, 9*(1), 20–31.

Connolly, M. (2019). Factors influencing consumer wine choice: The case of wine tourism. In *Management and marketing of wine tourism business: Theory, practice, and cases* (pp. 43–61). Palgrave Macmillan.

Estreicher, S. K. (2017). The beginning of wine and viticulture. *Physica Status Solidi C, 14*(7), 1700008.

Evans, J. R., & Mathur, A. (2005). The value of online surveys. *Internet Research, 15*(2), 195–219.

Ferguson, G., Cheah, I., & Lee, S. (2018). Preference for Australian premium wines in China: The effect of tourism. In *Food, wine and China* (pp. 164–188). Routledge.

Festa, G., Shams, S. R., Metallo, G., & Cuomo, M. T. (2020). Opportunities and challenges in the contribution of wine routes to wine tourism in Italy – A stakeholders' perspective of development. *Tourism Management Perspectives, 33*, 100585.

Figueroa, B. E., & Rotarou, E. S. (2018). Challenges and opportunities for the sustainable development of the wine tourism sector in Chile. *Journal of Wine Research, 29*(4), 243–264.

Ford, J. K., MacCallum, R. C., & Tait, M. (1986). The application of exploratory factor analysis in applied psychology: A critical review and analysis. *Personnel Psychology, 39*(2), 291–314.

Gu, Q., & Huang, S. (2019). Profiling Chinese wine tourists by wine tourism constraints: A comparison of Chinese Australians and long-haul Chinese tourists in Australia. *International Journal of Tourism Research, 21*(2), 206–220.

Gu, Q., Qiu Zhang, H., King, B., & Huang, S. (2018). Wine tourism involvement: A segmentation of Chinese tourists. *Journal of Travel & Tourism Marketing, 35*(5), 633–648.

Hair, J. F., Black, W. C., Babin, B. J., Anderson, R. E., & Tatham, R. L. (2014). *Pearson new international edition. Multivariate data analysis* (7th ed.). Pearson Education Limited Harlow.

Hall, C. M., Johnson, G., Cambourne, B., Macionis, N., Mitchell, R., & Sharples, L. (2002). Wine tourism: An introduction. In wine tourism around the world. In C. M. Hall, L. Sharples, B. Cambourne, & N. Macionis (Eds.), (1st ed., pp. 1–23). Routledge. ISBN-978 0 75065 466 1.

Hall, C. M., Mitchell, R., & Sharples, L. (2004). Consuming places the role of food, wine and tourism in regional development: The role of food, wine and tourism in regional development. In *Food tourism around the world* (pp. 25–59). Routledge.

Hernández, A. L., Alarcón, S., & Meraz Ruiz, L. (2022). Segmentation of wine tourism experience in Mexican wine regions using netnography. *International Journal of Wine Business Research, 34*(3), 427–446.

Kaiser, H. F. (1958). The varimax criterion for analytic rotation in factor analysis. *Psychometrika, 23*(3), 187–200.

Karampatea, A., Mamalis, S., Bouloumpasi, E., Skendi, A., & Kamenidou, I. (2024). Wine and gastronomic tourism in the drama region. *Tourism and Hospitality, 5*(3), 625–638.

Kastenholz, E., Paço, A., & Nave, A. (2023). Wine tourism in rural areas – Hopes and fears amongst local residents. *Worldwide Hospitality and Tourism Themes, 15*(1), 29–40.

Kelley, K. M., Bruwer, J., Zelinskie, J., Gardner, D. M., Govindasamy, R., Hyde, J., & Rickard, B. J. (2019). Travel group member type effects in wine tourism: An ECHAID segmentation. *Tourism Recreation Research, 44*(1), 54–65.

Kim, A. K., Del Chiappa, G., & Napolitano, E. (2019). Understanding the wine tourist markets' motivations, travel constraints and perceptions of destination attributes: A case study of winery visitors in Sardinia, Italy. In *Wine tourism destination management and marketing: Theory and cases* (pp. 9–27). Springer.

Koksal, M. H. (2021). Segmentation of wine consumers based on level of involvement: A case of Lebanon. *British Food Journal, 123*(3), 926–942.

Kruger, M., & Viljoen, A. (2022). That old saying about wine and age: Identifying south African age-cohort preferences. *International Journal of Wine Business Research, 34*(4), 495–522.

Mamalis, S., Kamenidou, I., & Fourkioti, G. (2023, August). Winery features considered significant for visiting choice. In *International conference of the International Association of Cultural and Digital Tourism* (pp. 891–900). Springer Nature Switzerland.

Marujo, N., Serra, J., Lima, J., & do Rosário Borges, M. (2022). Improving the tourist experience at world heritage sites: How to explore motivations using expenditure-based segmentation. In *Handbook on the tourist experience* (pp. 102–117). Edward Elgar Publishing.

Masberg, Barbara, Wine Tourism Best Practices (2014). Symposium of University research and creative expression (SOURCE). 141. https://digitalcommons.cwu.edu/source/2014/oralpresentations/141

Napolitano, E., Atzeni, M., Kim, A., & Del Chiappa, G. (2022). Diverse socializing patterns in wine tourist experiences: A segmentation-based analysis of visitors to the wineries in South Australia. *International Journal of Tourism Research, 24*(6), 839–853.

Nella, A., & Christou, E. (2014). Segmenting wine tourists on the basis of involvement with wine. *Journal of Travel & Tourism Marketing, 31*(7), 783–798.

Nella, A., & Christou, E. (2021). Market segmentation for wine tourism: Identifying sub-groups of winery visitors. *European Journal of Tourism Research, 29*, 2903–2903.

Payini, V., Bolar, K., Mallya, J., & Kamath, V. (2022). Modeling hedonic motive–based segments of wine festival visitors using decision tree approach. *International Journal of Wine Business Research, 34*(1), 19–36.

Pitoska, E., Theodorli, E., & Altini, A. (2023, June). Entrepreneurship of winemaking Enterprises in Mountain Less-Favored Areas: An empirical study. In *International conference on applied economics* (pp. 1113–1126). Springer Nature Switzerland.
Saayman, M., Saayman, A., & Joubert, E. M. (2012). Expenditure-based segmentation of visitors to the wacky wine festival. *Tourism Recreation Research, 37*(3), 215–225.
Santos, V., Ramos, P., Almeida, N., Marôco, J., & Santos-Pavón, E. (2020). Wine tourist profiling in the Porto wine cellars: Segmentation based on wine product involvement. *Anatolia, 31*(4), 577–590.
Santos, V., Ramos, P., Sousa, B., & Valeri, M. (2022). Towards a framework for the global wine tourism system. *Journal of Organizational Change Management, 35*(2), 348–360.
Seyedimany, A., & Koksal, M. H. (2022). Segmentation of Turkish wine consumers based on generational cohorts: An exploratory study. *Sustainability, 14*(5), 3031.
Simpson, K., & Bretherton, P. (2004). Market segmentation by consumer lifestyle in a wine tourism setting. In *Proceedings of the New Zealand tourism and hospitality research conference* (pp. 377–385).
Simpson, K., Bretherton, P., & de Vere, G. (2005). Lifestyle market segmentation, small business entrepreneurs, and the New Zealand wine tourism industry. *Journal of Quality Assurance in Hospitality & Tourism, 5*(2–4), 157–188.
Stergiou, D. P., Airey, D., & Apostolakis, A. (2018). The winery experience from the perspective of generation Z. *International Journal of Wine Business Research, 30*(2), 169–184.
Sun, Y. Y., & Drakeman, D. (2022). The double-edged sword of wine tourism: The economic and environmental impacts of wine tourism in Australia. *Journal of Sustainable Tourism, 30*(4), 932–949.
The wine lab.eu. (2018). *UNIT 1 – Wine tourism: definition, opportunities, implications and strategies*. https://training.thewinelab.eu/
Trigo, A., & Silva, P. (2022). Sustainable development directions for wine tourism in Douro wine region, Portugal. *Sustainability, 14*(7), 3949.
Vîlcea, C., Licurici, M., & Popescu, L. (2024). The role of websites in promoting wine tourism: An evaluation of Romanian wineries. *Sustainability, 16*(15), 6336.
Vorobiova, N., Pinto, P., Pintassilgo, P., & Lavandoski, J. (2020) Motivations of tourists in wine regions: The case of La Rioja, Spain. *International Journal of Wine Business Research, 32*(3), 353–371.

Chapter 4
Hydrological Balance Assessment for Sustainable Water Resources Management: Case Studies in Ghana, Burkina Faso, Kenya, and Tanzania

Konstadinos Mattas, Pantazis Georgiou, Christos Mattas, Dimitra Lazaridou, Stefanos Nastis, Seddaiu Giovanna, Mamadou Traore, Kombiok James Mantent, Sheila Okoth, and Deodatus Kiriba

4.1 Introduction

Sustainable water resources management is essential for addressing climate change, population growth, and increasing agricultural demands. Hydrological studies often involve developing models to simulate water behavior under changing climatic

K. Mattas · S. Nastis
Department of Agricultural Economics, Aristotle University of Thessaloniki, Thessaloniki, Greece
e-mail: mattas@auth.gr; snastis@auth.gr

P. Georgiou (✉)
Department of Hydraulics, Soil Science and Agricultural Engineering, School of Agriculture, Aristotle University of Thessaloniki, Thessaloniki, Greece
e-mail: pantaz@agro.auth.gr

C. Mattas
Department of Structural, Historical and Applied Geology, School of Geology, Aristotle University of Thessaloniki, Thessaloniki, Greece
e-mail: cmattas@geo.auth.gr

D. Lazaridou
Department of Agricultural Economics, School of Agriculture, Aristotle University of Thessaloniki, Thessaloniki, Greece

Department of Forestry and Natural Environment Management, Agricultural University of Athens, Athens, Greece
e-mail: dlazaridou@aua.gr

S. Giovanna
Department of Agricultural Sciences, University of Sassari, Sassari, Italy
e-mail: gseddaiu@uniss.it

K. Mattas et al. (eds.), *Biodiversity in Agri-Food Systems*, Cooperative Management, https://doi.org/10.1007/978-3-032-06385-4_4

conditions and human activities, aiding in decision-making for stakeholders, managers, and policymakers. Calculating a reliable water balance is a fundamental yet challenging step in these studies, as its accuracy depends on the quality and density of available data and the precision of hydrogeological parameter estimations.

Within the EWA-BELT project framework, the Thornthwaite-Mather method (Thornthwaite & Mather, 1955, 1957) was applied to estimate the water balance in selected watersheds across Ghana, Burkina Faso, Kenya, and Tanzania. This method is particularly useful in regions with limited hydrological monitoring as it highlights seasonal variations in precipitation, evapotranspiration, soil moisture, irrigation requirements, and runoff (Donker, 1987).

Evapotranspiration is crucial in the water balance and is influenced by meteorological conditions, soil properties, and vegetation characteristics. Estimating actual evapotranspiration in field conditions is complex; hence, potential evapotranspiration is first calculated and subsequently used to determine actual evapotranspiration via the water balance method (Burak et al., 2022).

This widely used method provides a framework for understanding precipitation, evapotranspiration, and soil water storage interactions over time, helping to determine whether a region experiences water surplus, deficit, or equilibrium from a hydrological perspective. The necessary input data—including digital elevation models, land use, soil type, geological conditions, precipitation, and temperature—were derived from an extensive literature review and online sources. The Thornthwaite-Mather method is particularly useful in the early stages of hydrological planning (Hendrayana et al., 2021).

This study aims to evaluate the hydrological balance of water resources in four selected river basins using the Thornthwaite-Mather method. The specific objectives include: (a) Quantifying water balance components—precipitation, potential and actual evapotranspiration, runoff, and deep percolation; (b) Assessing seasonal and spatial variations in water availability, highlighting critical periods of water surplus and deficit; (c) Identifying key factors influencing hydrological dynamics, such as climate, land use, and geology; and (d) Providing recommendations for sustainable water resource management based on findings from the EWA-BELT project case studies.

M. Traore
University Nazi BONI, Bobo-Dioulasso, Burkina Faso

K. J. Mantent
Kundok Development Consult (KDC), Tamale, Ghana

S. Okoth
Department of Biology, Faculty of Science and Technology, University of Nairobi, Nairobi, Kenya

D. Kiriba
Tanzania Agricultural Research Institute (TARI)-Selian Centre, Arusha, Tanzania

4.2 Description of the Methodology

The Thornthwaite-Mather method is widely used due to its simplicity, ease of application, and effectiveness in assessing evapotranspiration and water balance across various climates and landscapes (Karpouzos et al., 2011; Koukouli et al., 2022; Mammoliti et al., 2021). Its reliance on readily available climatic data makes it preferable in hydrological modeling studies (Burak et al., 2022). The method estimates water balance using soil moisture capacity and accounts for key parameters, including precipitation (P), potential evapotranspiration (PET), accumulated potential water loss (APWL), soil moisture (SM), deficit (DEF), surplus (SUR), actual evapotranspiration (AET), deep percolation (I), and runoff (RO).

Figure 4.1 represents the mechanisms behind the occurrence of runoff based on Thornthwaite's water balance approach.

The method primarily requires mean monthly temperature and precipitation data, making it applicable in diverse geographical locations, particularly where hydrological or climatic data are scarce. It computes potential evapotranspiration (PET), which represents the amount of water that would be evaporated and transpired if sufficient water were available. The method also calculates actual evapotranspiration (AET), soil moisture variations, water surplus, and deficit, offering a straightforward and widely accepted approach for evaluating water balance in hydrology and climate studies.

The equation for estimating PET on a monthly basis is:

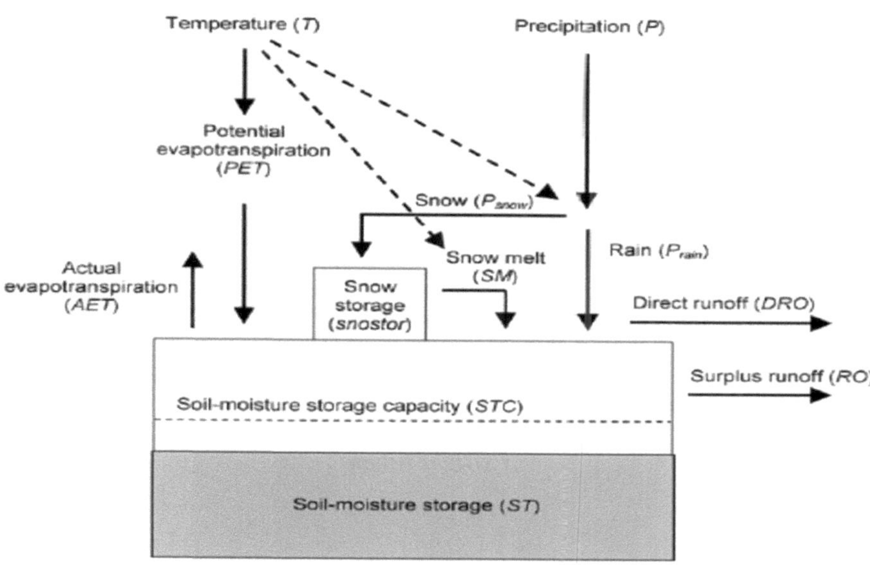

Fig. 4.1 Diagram of Thornthwaite-Mather's water balance method (McCabe & Markstrom, 2007)

$$PET_i = 16b_i \left(10\frac{T_i}{I}\right)^{\alpha} \tag{4.1}$$

where: PET = potential evapotranspiration (mm), b = radiation parameter for a specific latitude, T = mean air temperature (°C), I = annual heat index, α = complex function of heat index, and i = monthly index.

The heat index I is calculated from the mean monthly temperatures and reflects the climatic characteristics of the region. The formula is:

$$\mathrm{I} = \sum_{i=1}^{12}\left(\frac{T_i}{5}\right)^{1.514} \tag{4.2}$$

where: T_i is the mean monthly temperature for month i (in °C).

The exponent α is a function of the heat index I, and it is calculated using the following empirical formula:

$$\alpha = 6.75\cdot10^{-7}I^3 - 7.71\cdot10^{-5}I^2 + 1.792\cdot10^{-2}I + 0.49239 \tag{4.3}$$

The water balance is calculated using the method formulated by Thornthwaite and Mather (1957), Alley (1984), and Steenhuis and Van Der Molen (1986):

- Surplus of the water budget for month i [SUR_i]

$$\text{If } P_i - PET_i > 0 \text{ then } SUR_i = P_i - PET_i \tag{4.4}$$

$$\text{If } P_i - PET_i \le 0 \text{ then } SUR_i = 0 \tag{4.5}$$

- Deficit of the water budget for month i [DEF_i]

$$\text{If } P_i - PET_i < 0 \text{ then } DEF_i = \left|P_i - PET_i\right| \tag{4.6}$$

$$\text{If } P_i - PET_i \ge 0 \text{ then } DEF_i = 0 \tag{4.7}$$

- Accumulated Potential Water Loss for month i [$APWL_i$]

$$\text{If } DEF_i \ne 0 \text{ then } APWL_i = APWL_{i-1} + DEF_i \tag{4.8}$$

$$\text{If } SUR_i \ne 0 \text{ then}$$

$$APWL_i = -S_o \ln\left\{\frac{S_{i-1} + SUR_i}{S_{\max}}\right\} \text{when } S_{i-1} + SUR_i < S_{\max} \tag{4.9}$$

$$APWL_i = 0 \text{ when } S_{i-1} + SUR_i \geq S_{max} \tag{4.10}$$

where: S_{max} = maximum soil moisture holding capacity.

- Soil moisture for month i [S_i]

$$\text{If } DEF_i \neq 0 \text{ then } S_i = S_{max} e^{-\frac{APWL_i}{S_{max}}} \tag{4.11}$$

$$\text{If } SUR_i \neq 0 \text{ then } S_i = S_{i-1} + SUR_i \tag{4.12}$$

- Soil moisture consumption for month i [ΔS_i]

$$\Delta S_i = S_{i-1} - S_i \text{ when } S_{i-1} - S_i > 0 \tag{4.13}$$

$$\Delta S_i = 0 \text{ when } S_{i-1} - S_i \leq 0 \tag{4.14}$$

- Actual evapotranspiration for month i [AET]$_i$

$$AET_i = PET_i \text{ when } \Delta S_i = 0 \tag{4.15}$$

$$AET_i = P_i + \Delta S_i \text{ when } \Delta S_i > 0 \tag{4.16}$$

- Volume for runoff for month i [VS_i]

$$VS_i = 0 \text{ when } SN_i < 0.0 \tag{4.17}$$

$$VS_i = SN_i - S_i \text{ when } SN_i \geq 0.0 \tag{4.18}$$

where: $SN_i = S_{i-1} - DEF_i$ or $SN_i = S_{i-1} + SUR_i$

- Deep percolation for month i [I_i]

$$I_i = b \cdot VS_i \tag{4.19}$$

where: b = deep percolation coefficient (%).

- Runoff for month i [RO_i]

$$RO_i = d \cdot (1-b) VS_i + (1-d) DVO_i \tag{4.20}$$

where: d = lag coefficient of runoff (%), DVO_i = DVF_{i-1} and DVF_i = d DVO_i + (1 – a) VS_i

All methods include simplifications of reality and come with inherent constraints. The quality and availability of input data also play a significant role in determining the reliability of the method and the accuracy of the results. The Thornthwaite-Mather method presents the following limitations:

- It only considers the temperature for calculating PET and does not take into account other climate variables like wind, humidity, or solar radiation.
- Soil moisture storage capacity (S_{max}) is often assumed and might not reflect complex soil properties accurately.

Despite its limitations, the method is widely used because of its simplicity and ability to provide insight into the seasonal and annual water balance of a basin using minimal climatic data, mainly temperature and precipitation.

4.3 Study Areas and Data Sets

The Thornthwaite-Mather method was applied to selected river basins (case studies) in Ghana, Burkina Faso, Kenya, and Tanzania to determine the water balance in each one of them. For the application of the method and the assessment of the water balance, geological and hydrometeorological (temperature and precipitation) data were used, as well as the land use data of the areas chosen. Land use data and GIS files were acquired from https://data.apps.fao.org/ (last accessed 02/25/2025). Hydrogeology maps were derived by https://earthwise.bgs.ac.uk/ (last accessed 02/25/2025). The meteorological data was collected from https://aquastat.fao.org/climate-information-tool/ (last accessed 02/25/2025).

Land use significantly impacts hydrological processes, influencing infiltration, runoff, and water quality. Human activities, such as agriculture and urbanization, alter natural water flow patterns and affect groundwater recharge. Geological conditions also play a crucial role in determining water infiltration, storage, and quality.

Meteorological conditions are vital in agriculture, disaster management, urban planning, meteorological studies, and energy production. Therefore, meteorological data is the most important type of data in hydrological studies. They are used in hydrological models, natural resources management studies, forecasting models, climate pattern studies, etc. They can be collected from weather stations, satellites, and weather balloons, providing real-time and historical information. The minimum data requirements for calculating the water balance of a catchment are air temperature, precipitation height, and initial soil moisture. This study examines meteorological data, including monthly temperature and precipitation, for the period from 2000 to 2021. The following paragraphs provide a brief description of each case study area.

Ghana Case Study

Ghana is located in the Gulf of Guinea in West Africa. The selected river basin in Ghana is situated in the Upper East Region of the country, as shown in Fig. 4.2, and its area was calculated as equal to 1895.90 km^2.

The geomorphological features of the river basin are presented in Table 4.1.

The data for the Ghana case study land use was classified into nine classes. The corresponding areas of the land use categories are shown in Table 4.2, and their spatial distribution is illustrated on the map of Fig. 4.3.

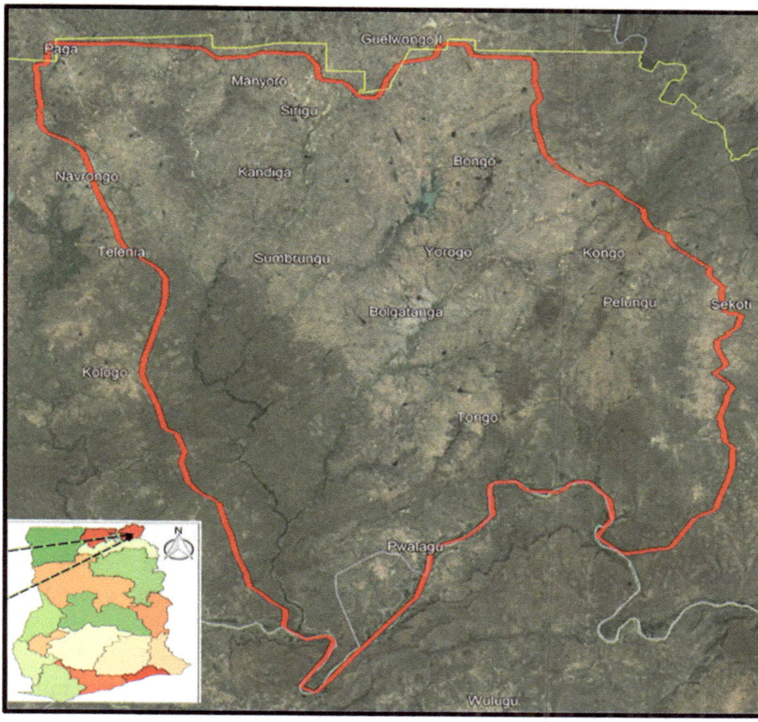

Fig. 4.2 Ghana case study boundaries (Google Earth image)

Table 4.1 River basin geomorphological features of Ghana case study

Basin characteristic	
Area (km²)	1895.9
Mean elevation (m)	191
Max elevation (m)	394
Min elevation (m)	120

The Ghana case study area is mainly occupied by Precambrian rocks as depicted in Table 4.3 and Fig. 4.4. Precambrian rocks are mainly metamorphic, plutonic, and volcanic.

Based on the geological features and land use in the Ghana case study along with the literature review (Acheampong, 1988; Liebe et al , 2009), the deep percolation coefficient is estimated as being equal to 5%, the lag coefficient of runoff is equal to 30% and the maximum soil moisture holding capacity is equal to 70 mm.

The average monthly temperature values and the average monthly precipitation for the period 2000–2021 are given in Table 4.4.

The values of the parameters of the Thornthwaite-Mather method for the Ghana case study are given in Table 4.5.

Table 4.2 Land use distribution in Ghana case study area

Land use	Area (km^2)
Cultivated and managed terrestrial area(s)/natural and semi-natural primarily terrestrial vegetation	755.92
Natural and semi-natural primarily terrestrial vegetation/cultivated and managed terrestrial area(s)	64.95
Artificial surfaces and associated area(s)	2.05
Bare area(s)	0.37
Rainfed herbaceous crop(s)	367.83
Broadleaved deciduous (40–(20–10)%) woodland	28.76
Closed to open trees/closed to open (100–40)% Shrubland (thicket)/ herbaceous closed to open vegetation	68.12
Closed to open Shrubland (thicket)	603.04
Artificial waterbodies/natural waterbodies	4.85

Burkina Faso

Burkina Faso is located in West Africa, south of the Sahara Desert. The selected river basin administratively belongs to the Bereba District in the Province of Tuy in the Region of Hauts-Bassins, as shown in Fig. 4.5, and its extent was calculated to be equal to 12,223.52 km^2.

The river basin's geomorphological features are shown in Table 4.5.

The land use data for the Burkina Faso case study was classified into nine classes. Land use categories for the corresponding areas are shown in Table 4.6, and their spatial distribution is illustrated in Fig. 4.6.

Burkina Faso's geology consists of Proterozoic to Paleozoic (meta)sedimentary rocks at a percentage of 96.37%. They mostly include sedimentary and metamorphic rocks that formed between the Proterozoic and Paleozoic eras. The geological features of the case study area are shown in Table 4.7, and the geological map is illustrated in Fig. 4.7.

Based on the geological features, the land use in the Burkina Faso case study, and the literature review (Müller et al., 2018), the deep percolation coefficient is estimated to be equal to 15%, the lag coefficient of runoff is equal to 30% and the maximum soil moisture holding capacity is equal to 80 mm.

The average monthly temperature values and the average monthly precipitation for the period 2000–2021 are given in Table 4.8.

Kenya Case Study

Kenya is located on the East Coast of Africa, bordering the Indian Ocean to the east, Sudan and Ethiopia to the north, Somalia to the north-east, Uganda to the west, and Tanzania to the south. The selected river basin in the Kenya area includes part of the Nyakach Region, Kisumu County, as shown in Fig. 4.8, and its area was calculated to be 3331.73 km^2.

The river basin's geomorphological features are shown in Table 4.9.

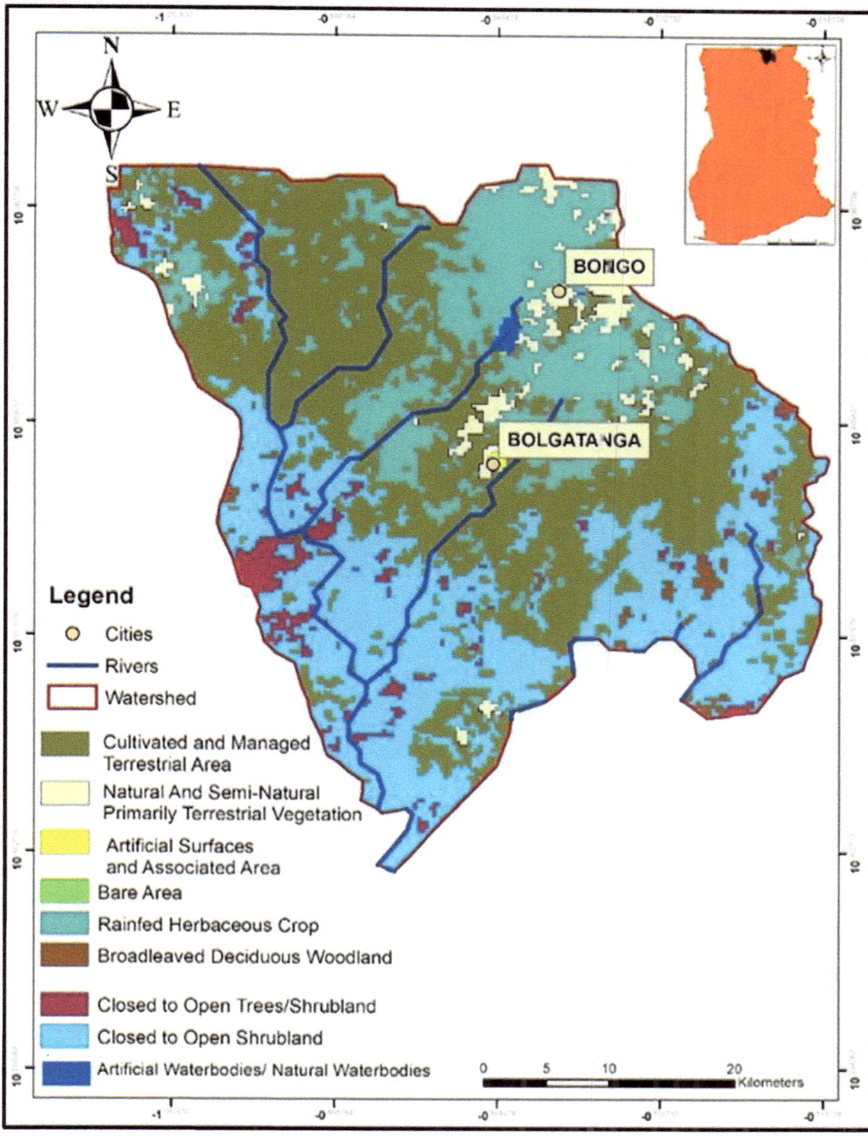

Fig. 4.3 Land use map of Ghana case study

Table 4.3 Geological features of Ghana case study area

Geology	Area (km²)
Precambrian craton	1881.38
Sedimentary—Proterozoic-Palaeozoic—Voltaian	14.52

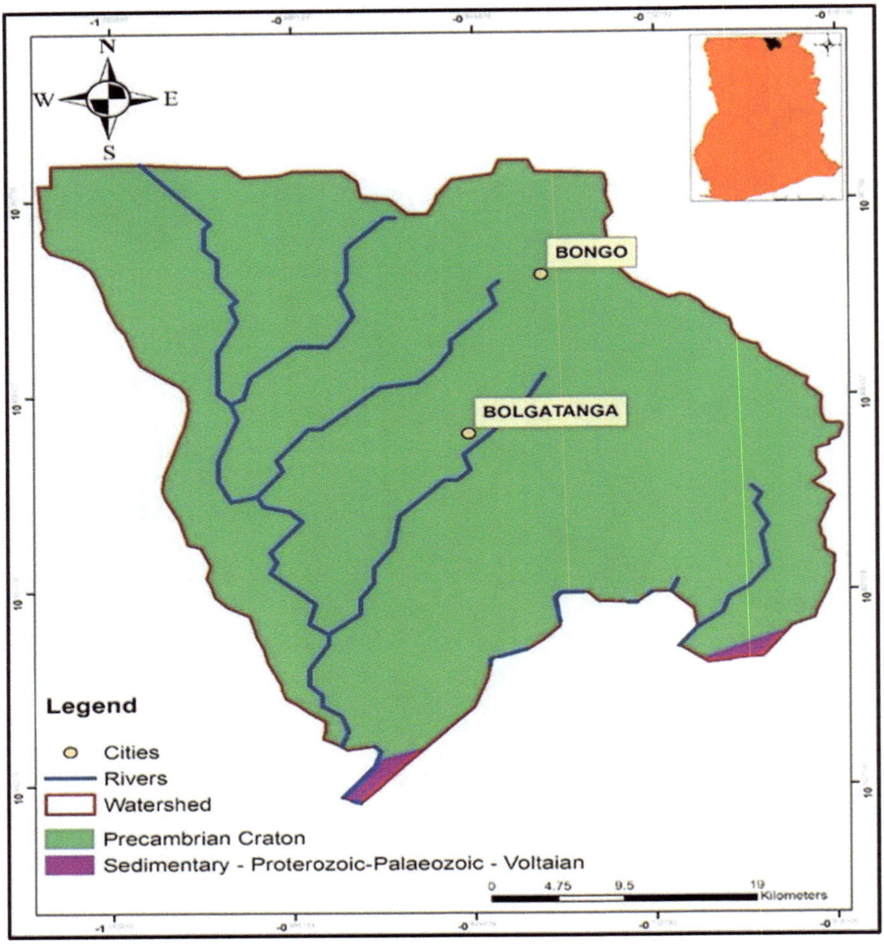

Fig. 4.4 Geological map of Ghana case study

Land use data was classified into thirteen categories. The specific areas for each category are presented in Table 4.10, and the land use map of Kenya is shown in Fig. 4.9.

Volcanic rocks cover 55.48% of the total case study area. They are a subset of igneous rocks formed from the solidification of molten rock material that erupts from a volcano. Table 4.11 presents all the types of geological formations that occur in the Kenya case study and their distribution, and the related map is illustrated in Fig. 4.10.

Table 4.4 Mean monthly precipitation and temperature in the case study of Ghana for the years 2000–2021

	J	F	M	A	M	J	J	A	S	O	N	D	Mean annual
Mean monthly precipitation (mm)	0.3	2.6	9.6	31.2	57.9	71.8	167.6	238.5	173.9	62.9	3.7	0.0	820
Mean monthly temperature (°C)	27.3	29.6	31.9	32.4	31.1	29.0	26.9	26.0	26.3	27.6	28.2	27.3	28.6

Table 4.5 River basin geomorphological features of Burkina Faso case study

Basin characteristic	
Area (km²)	12278.5
Mean elevation (m)	370.4
Max elevation (m)	693
Min elevation (m)	262

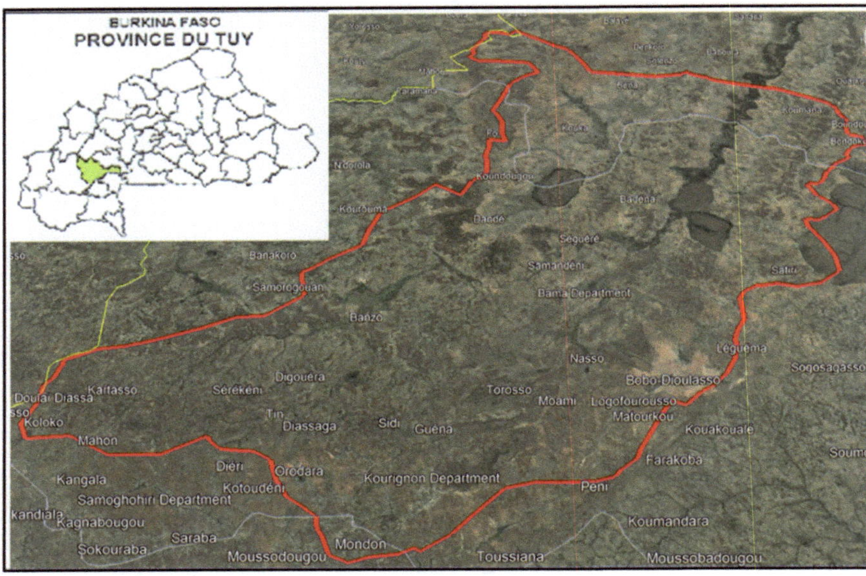

Fig. 4.5 Burkina Faso case study boundaries. (Google Earth image)

Table 4.6 Land use distribution in Burkina Faso case study area

Land use	Area (km²)
Cultivated and managed terrestrial area(s)/natural and semi-natural primarily terrestrial vegetation	5582.94
Natural and semi-natural primarily terrestrial vegetation/cultivated and managed terrestrial area(s)	1650.90
Artificial surfaces and associated area(s)	20.59
Bare area(s)	0.93
Rainfed herbaceous crop(s)	2893.12
Sparse Woody vegetation/herbaceous sparse vegetation	5.96
Broadleaved deciduous (40–(20–10)%) woodland	61.30
Closed to open trees//closed to open (100–40)% Shrubland (thicket)/herbaceous closed to open vegetation	190.19
Closed to open Shrubland (thicket)	1813.86
Herbaceous closed to open vegetation/closed to open trees/closed to open Shrubland (thicket)	0.93
Artificial waterbodies//natural waterbodies	2.79

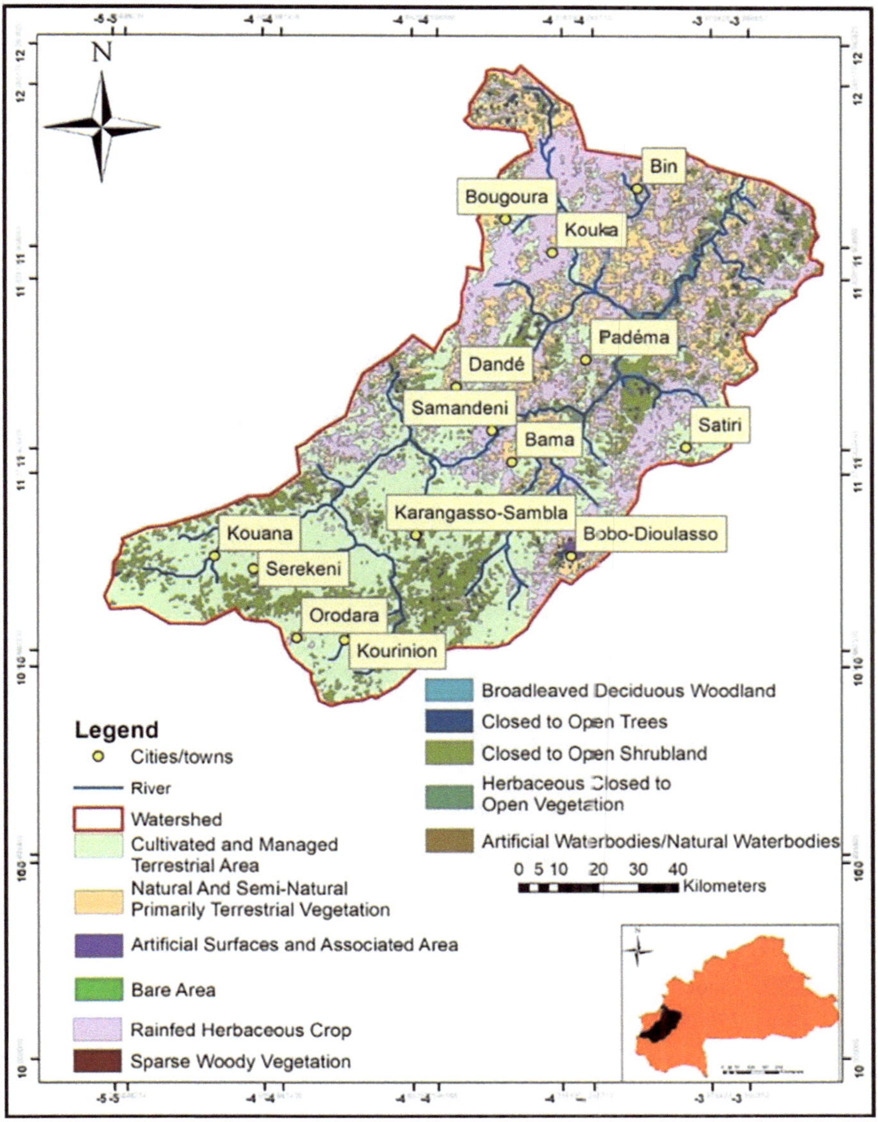

Fig. 4.6 Land use map of the Burkina Faso case study

Based on the geological features and land use in the Kenya case study along with thr literature review (Castelino & Khamala, 1979), the deep percolation coefficient is estimated as being equal to 9%, the lag coefficient of the runoff is equal to 30% and the maximum soil moisture holding capacity is equal to 100 mm.

The average monthly temperature values and the average monthly precipitation for the period 2000–2021 are given in Table 4.12.

Table 4.7 Geological features of the Burkina Faso case study area

Geology	Area (km^2)
Proterozoic to Palaeozoic (meta)sedimentary	11779.87
Igneous intrusive—dolerite	443.67

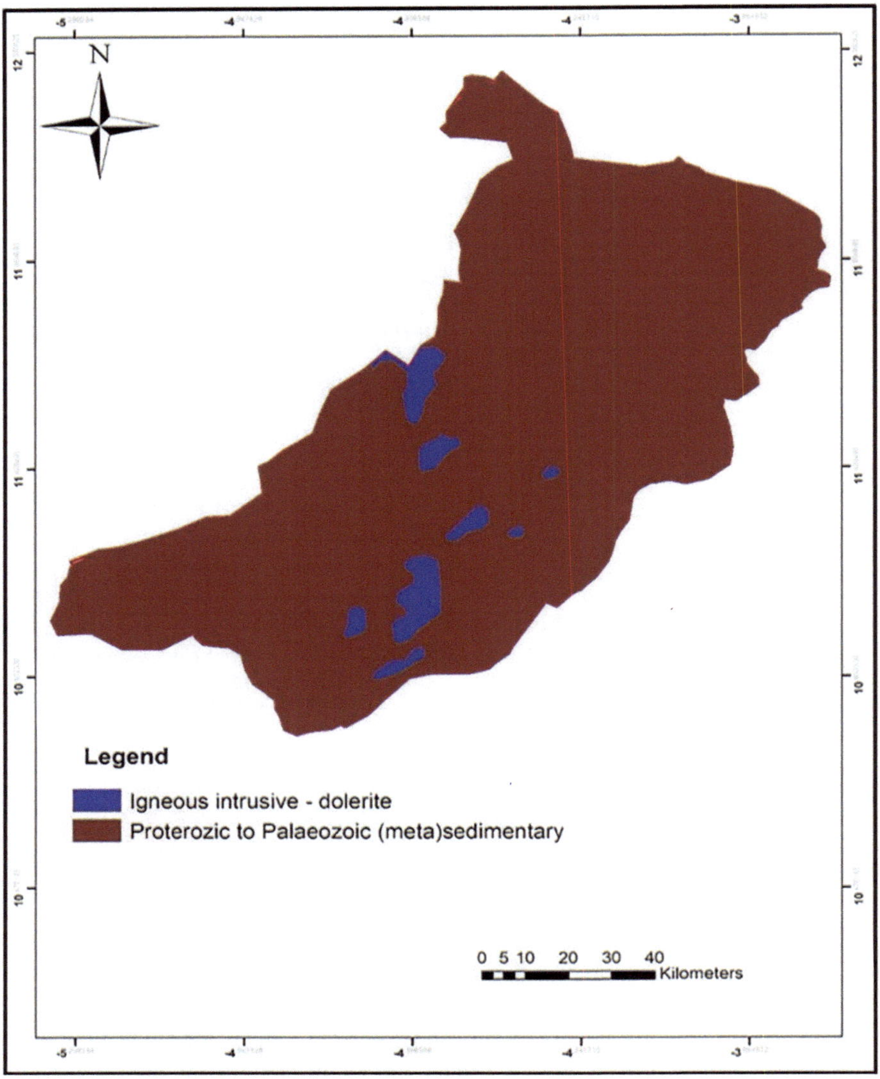

Fig. 4.7 Geological map of the Burkina Faso case study

Table 4.8 Mean monthly precipitation and temperature in the case study of Burkina Faso for the years 2000–2021

	J	F	M	A	M	J	J	A	S	O	N	D	Mean annual
Mean monthly precipitation (mm)	0.6	0.5	5.3	22.1	56.6	83.9	178.5	245.4	152.4	52.1	2.6	0.0	800
Mean monthly temperature (°C)	26.0	28.5	30.9	32.0	30.9	28.9	26.7	25.7	26.1	27.5	27.5	26.2	28.1

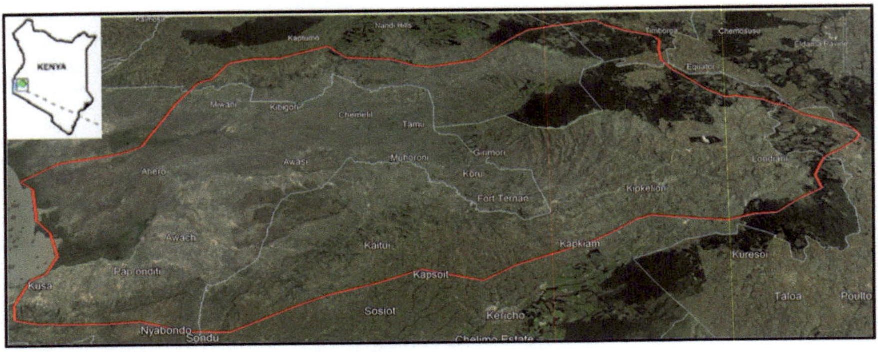

Fig. 4.8 Kenya case study boundaries. (Google Earth image)

Table 4.9 River basin geomorphological features of Kenya case study

Basin characteristic	
Area (km²)	3331.7
Mean elevation (m)	1690.6
Max elevation (m)	2884
Min elevation (m)	1132

Table 4.10 Land use distribution in the Kenya case study area

Land Use	Area (km²)
Cultivated and managed terrestrial area(s)/natural and semi-natural primarily terrestrial vegetation	1418.25
Natural and semi-natural primarily terrestrial vegetation/cultivated and managed terrestrial area(s)	684.34
Artificial surfaces and associated area(s)	2.67
Bare area(s)	3.05
Rainfed herbaceous crop(s)	390.54
Broadleaved deciduous (40–(20–10)%) woodland	74.25
Needle leaved deciduous (40–(20–10)%) woodland/needle leaved Evergreen (40–(20–10)%) woodland	1.88
Closed to open trees/closed to open (100–40)% Shrubland (thicket)/herbaceous closed to open vegetation	84.82
Closed to open Shrubland (thicket)	447.12
Herbaceous closed to open vegetation/closed to open trees/closed to open Shrubland (thicket)	0.95
Broadleaved evergreen closed to open trees/semi-deciduous closed to open trees	117.76
Broadleaved deciduous closed to open (100–40)% trees	100.95
Artificial waterbodies/natural waterbodies	5.15

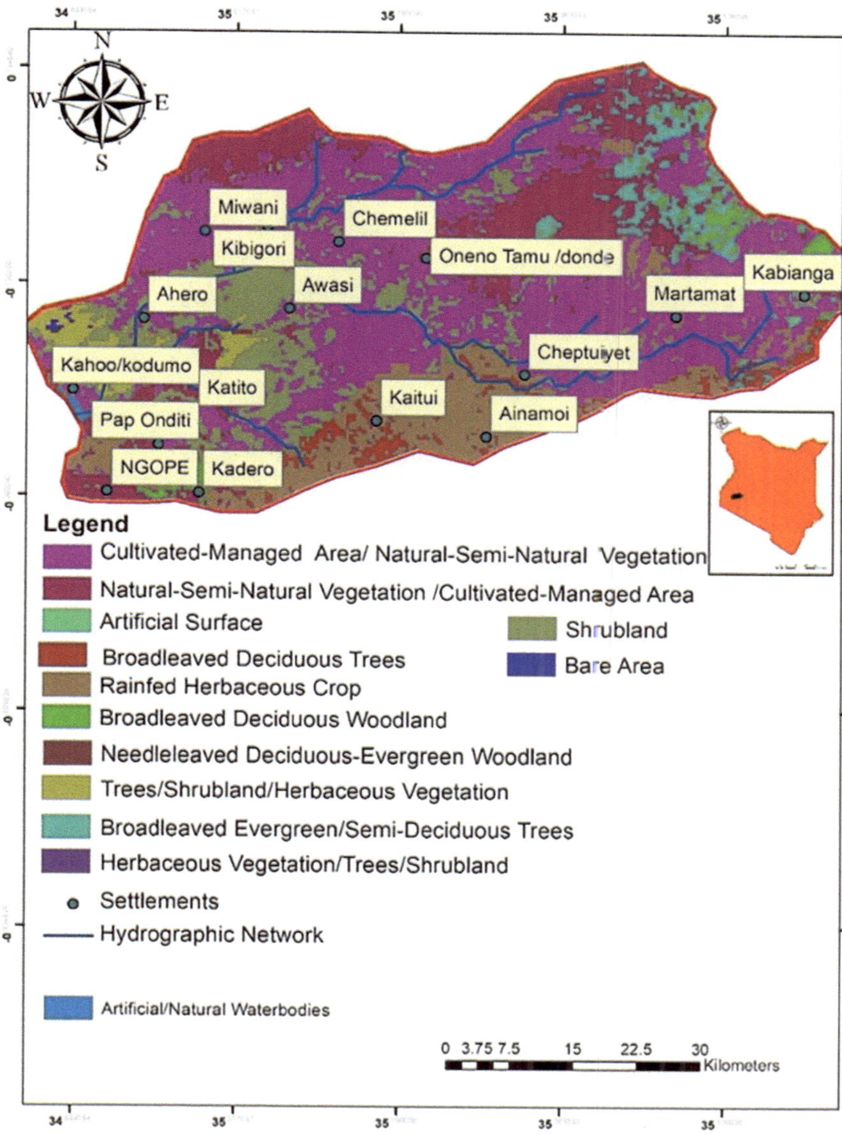

Fig. 4.9 Land use map in the Kenya case study

Table 4.11 Geological features of the Kenya case study area

Geology	Area (km²)
Precambrian/archaean/proterozoic	805.79
Igneous volcanic	1848.19
Sedimentary/tertiary/quaternary/unconsolidated	677.51

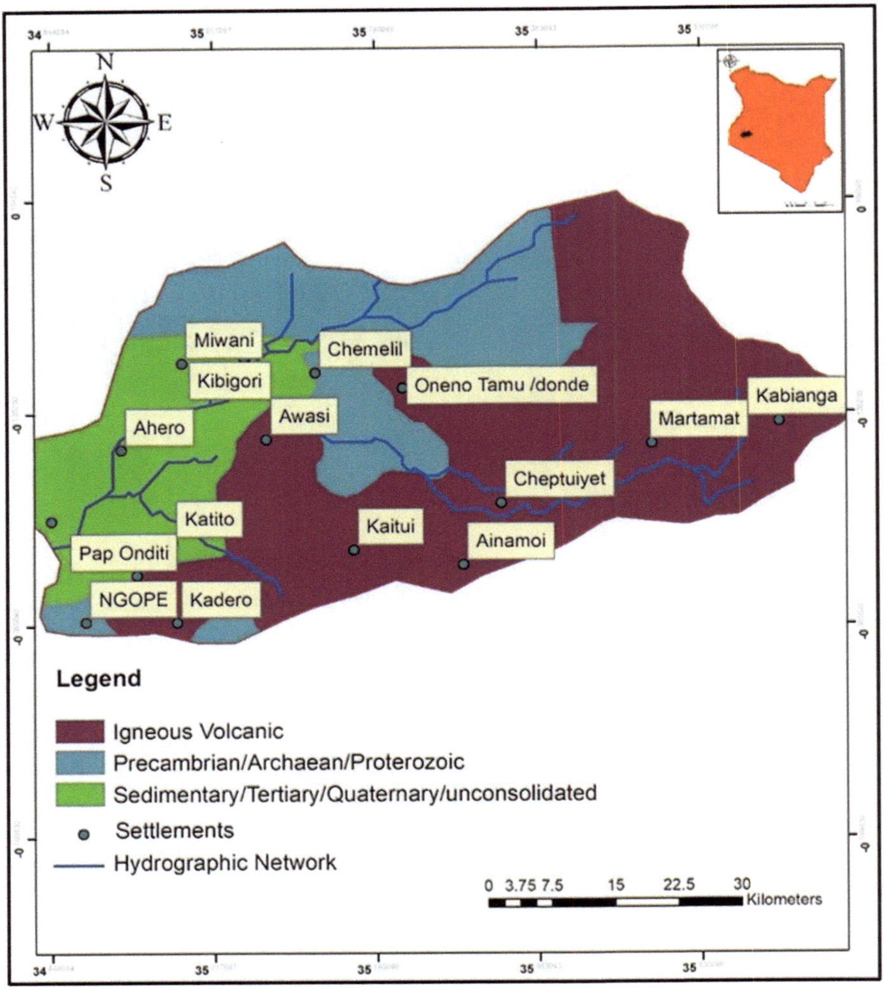

Fig. 4.10 Geological map of the Kenya case study

Tanzania Case Study

Tanzania is a country located in East Africa, bordering the Indian Ocean. The selected river basin in the Tanzania area administratively belongs to the Arusha Region in the Upper East Region of the country, as shown in Fig. 4.11, and its area was calculated to be 4699.07 km^2.

The geomorphological features in the river basin are depicted in Table 4.13.

Land use data for the selected case study basin in Tanzania was classified into sixteen classes. Their corresponding area is shown in Table 4.14, and the land use map of Tanzania is depicted in Fig. 4.12.

Table 4.12 Mean monthly precipitation and temperature in the case study of Kenya for the years 2000–2021

	J	F	M	A	M	J	J	A	S	O	N	D	Mean annual
Mean monthly precipitation (mm)	109.1	98.9	202.8	362.2	268.0	131.8	87.3	107.3	147.9	215.1	230.6	159.0	2120
Mean monthly temperature (°C)	20.9	21.6	21.3	20.1	19.4	19.0	18.9	19.4	19.9	20.3	20.2	20.4	20.1

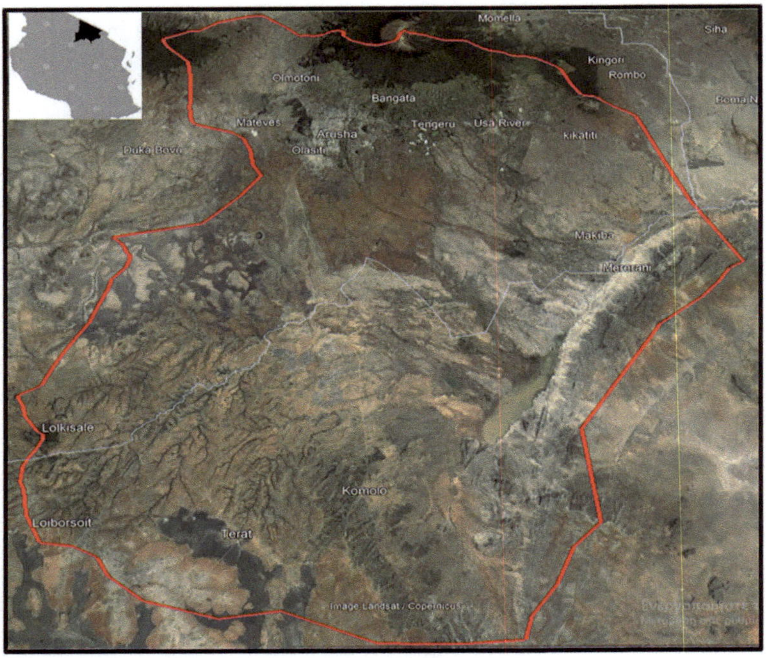

Fig. 4.11 Tanzania case study boundaries. (Google Earth image)

Table 4.13 River basin geomorphological features in the Tanzania case study

Basin characteristic	
Area (km²)	4699
Mean elevation (m)	1240.78
Max elevation (m)	4530
Min elevation (m)	870

Most of Tanzania's geological formations consist of Tertiary-Quaternary unconsolidated rocks, and they constitute approximately 59.69% of the total area. These rocks are often unconsolidated, meaning that they are not lithified into solid rock, and consist of loose materials. The categories that constitute Tanzania's geology and their distribution are presented in Table 4.15. The corresponding geology map is displayed in Fig. 4.13.

Based on the geological features and land use in the Tanzania case study along with the literature review (Fish, 2011), the deep percolation coefficient is estimated equal to 14%, the lag coefficient of the runoff is equal to 30%, and the maximum soil moisture holding capacity is estimated to be equal to 110 mm.

The average monthly temperature values and the average monthly precipitation for the period 2000–2021 are given in Table 4.16.

Table 4.14 Land use distribution in the Tanzania case study area

Land use	Area (km^2)
Natural and semi-natural primarily terrestrial vegetation//cultivated and managed terrestrial area(s)	1188.73
Artificial surfaces and associated area(s)	39.31
Rainfed herbaceous crop(s)	357.25
Open herbaceous vegetation	181.49
Herbaceous sparse vegetation	9.13
Broadleaved deciduous (40–(20–10)%) woodland	1013.52
Needleleaved deciduous (40–(20–10)%) woodland//needleleaved evergreen (40–(20–10)%) woodland	43.68
Closed to open trees//closed to open (100–40)% Shrubland (thicket)/herbaceous closed to open vegetation	116.84
Closed to open Shrubland (thicket)	548.25
Herbaceous closed to open vegetation//closed to open trees//closed to open shrubland (thicket)	74.98
Herbaceous closed to open vegetation//closed to open lichens/mosses	404.44
Broadleaved Evergreen closed to open trees//semi-deciduous closed to open trees	47.00
Broadleaved deciduous closed to open (100–40)% trees	580.01
Mixed closed to open trees	38.79
Closed to open (100–40)% closed to open (100–40)% broadleaved woodland on temporarily flooded land—water quality: Fresh water	54.95
Artificial waterbodies//natural waterbodies	0.68

4.4 Results and Discussion

The average water balance for each case study is presented in Tables 4.17, 4.18, 4.19, and 4.20. Figure 4.14 illustrates the water balance components for each case study area.

The four case studies illustrate distinct hydrological conditions influenced by geomorphology, climate, geology, and land use. Kenya receives the highest annual precipitation (2120 mm), followed by Tanzania (925 mm), Ghana (820 mm), and Burkina Faso (800 mm). It has been observed in many regions that precipitation increases with altitude. This phenomenon is known as the orographic enhancement of precipitation (Napoli et al., 2019). In the present study, the West Africa case studies reveal that, despite the Burkina Faso case study area being at a significantly higher elevation than the neighboring region in Ghana, it receives less precipitation. In East Africa, the precipitation in the Kenya case study, which is located at a much higher elevation, is significantly higher compared to the Tanzania case

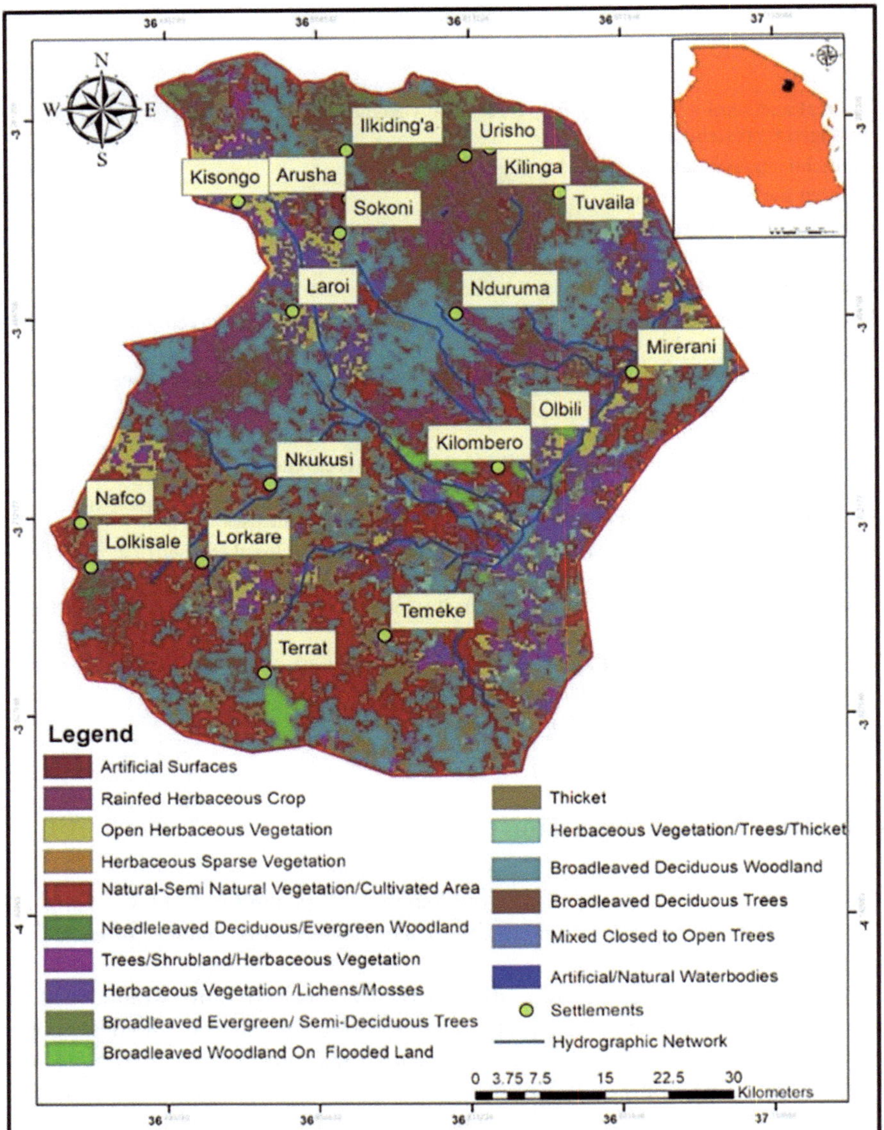

Fig. 4.12 Land use map of the Tanzania case study

Table 4.15 Geological features of the Tanzania case study area

Geology	Area (km^2)
Precambrian mobile/orogenic belt/ craton/	606.26
Tertiary-quaternary volcanic	1288.08
Tertiary-quaternary unconsolidated	2804.73

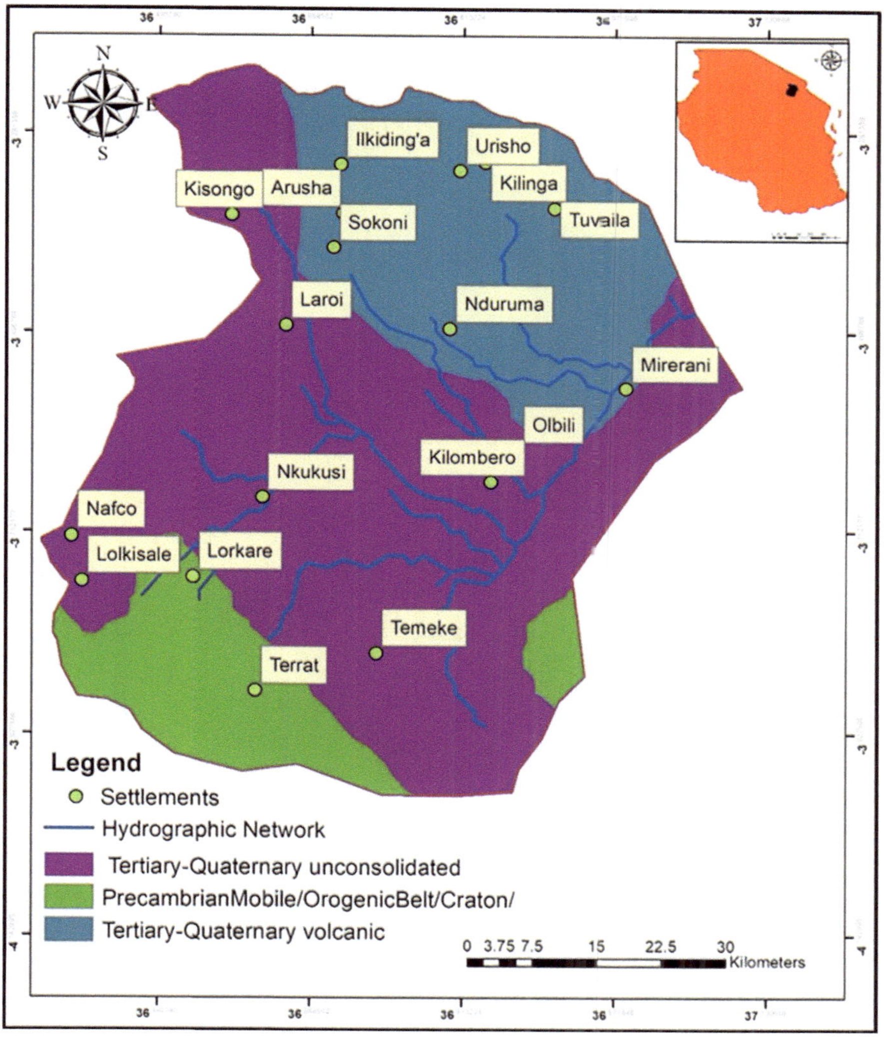

Fig. 4.13 Geological map of the Tanzania case study

Table 4.16 Mean monthly precipitation and temperature in Tanzania for the years 2000–2021

	J	F	M	A	M	J	J	A	S	O	N	D	Mean annual
Mean monthly precipitation (mm)	103.3	88.0	156.4	166.6	72.0	21.0	15.1	15.4	13.0	38.4	112.3	123.5	925
Mean monthly temperature (°C)	22.6	22.7	22.4	21.2	19.8	18.8	18.5	19.3	20.7	21.8	22.1	22.35	21.0

Table 4.17 Mean water balance in the Ghana case study

Precipitation (P) (mm)	**820**
Potential evapotranspiration (PET) (mm)	2249
Runoff (RO) (mm)	132 (16%)
Deep percolation (I) (mm)	7 (0.9%)
Actual evapotranspiration (AET) (mm)	681 (83.1%)
AET/PET*	30.3%

Table 4.18 Mean water balance in the Burkina Faso case study

Precipitation (P) (mm)	**800**
Potential evapotranspiration (PET) (mm)	2065
Runoff (RO) (mm)	110 (13.8%)
Deep percolation (I) (mm)	20 (2.5%)
Actual evapotranspiration (AET) (mm)	670 (83.7%)
AET/PET*	32.4%

Table 4.19 Mean water balance in the Kenya case study

Precipitation (P) (mm)	**2120**
Potential evapotranspiration (PET) (mm)	906
Runoff (RO) (mm)	1117 (52.7%)
Deep percolation (I) (mm)	111 (5.2%)
Actual evapotranspiration (AET) (mm)	892 (42.1%)
AET/PET*	98.5%

Table 4.20 Mean water balance in the Tanzania case study

Precipitation (P) (mm)	**925**
Potential evapotranspiration (PET) (mm)	956
Runoff (RO) (mm)	168 (18.2%)
Deep percolation (I) (mm)	27 (2.9%)
Actual evapotranspiration (AET) (mm)	730 (78.9%)
AET/PET[a]	76.4%

[a]Actual evapotranspiration/potential evapotranspiration

study.Temperature patterns show a clear trend: Burkina Faso and Ghana experience relatively higher mean annual temperatures (~28 °C) compared to Kenya (20.1 °C) and Tanzania (21 °C). The cooler climate of Kenya and Tanzania is likely due to their higher elevations, as reflected in their greater mean and maximum elevation values. The combination of the aforementioned parameters indicates that Ghana and Burkina Faso are more vulnerable to drought conditions.Kenya exhibits the highest runoff rate (1117 mm, 52.7%) compared to the other case study areas. To mitigate excessive water loss, Kenya would benefit from improved surface water conservation measures such as reservoir construction, rainwater harvesting, and reforestation. Meanwhile, Ghana and Burkina Faso, with lower runoff levels, could focus on soil moisture conservation strategies to support agriculture.Kenya (5.2%) has the highest deep percolation rate (favoring groundwater recharge), followed by Tanzania

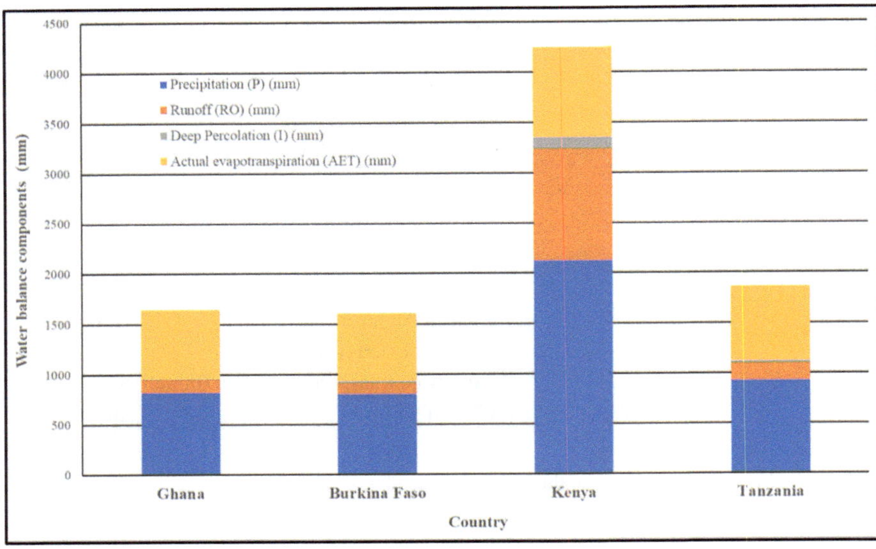

Fig. 4.14 Water balance components by country

(2.9%) and Burkina Faso (2.5%). Ghana's very low percolation rate (0.9%) is attributed to the low permeability of soils or compacted land. While Kenya and Tanzania could explore groundwater extraction for agricultural and domestic use, Ghana would benefit from improved soil management practices, such as reduced tillage and agroforestry, to enhance infiltration.

Burkina Faso and Ghana exhibit the highest actual evapotranspiration rates (83%–84%), meaning that most of their precipitation is lost as water vapor. Kenya has the lowest evapotranspiration rate (42.1%), likely due to cooler temperatures. Tanzania's evapotranspiration rate is moderately high (78.9%), which aligns with its climatic conditions. All four case studies indicate a water surplus, meaning that precipitation exceeds evapotranspiration. Kenya and Tanzania have the largest surpluses, suggesting greater potential for groundwater recharge or water storage for agricultural use. Despite having a surplus, Ghana and Burkina Faso have less water available for storage due to high evapotranspiration losses.

Kenya has the highest Actual Evapotranspiration to Potential Evapotranspiration (AET/PET) ratio (98.5%), indicating that almost all available moisture is utilized in evapotranspiration. Tanzania also has a high AET/PET ratio (76.4%), signifying efficient moisture utilization. In contrast, Ghana and Burkina Faso have lower AET/PET ratios (30.3% and 32.4%), suggesting that moisture demand exceeds by far the available water. High evaporative losses in Ghana and Burkina Faso necessitate water conservation strategies such as mulching and agroforestry. Kenya's low AET suggests that excess water is lost through runoff rather than being used efficiently.

Kenya and Tanzania exhibit high water surpluses, making them ideal for irrigation and water resource development. However, Kenya's high runoff underscores the need for better water retention strategies, such as reservoirs and infiltration trenches. In contrast, Ghana and Burkina Faso face significant evapotranspiration losses, making water conservation measures—such as mulching and rainwater harvesting—essential.

The findings have important policy and management implications for each region:

- **Ghana & Burkina Faso:** High evapotranspiration rates make these regions prone to droughts. Rainwater harvesting, irrigation efficiency, and soil moisture conservation (e.g., mulching, cover cropping) should be prioritized. The groundwater recharge rate is low, so alternative water storage solutions like surface reservoirs are needed.
- **Kenya:** Runoff management strategies (e.g., reservoirs, wetlands, infiltration trenches) should be implemented to retain more water for irrigation and drinking water supply. Soil conservation methods, such as agroforestry and terracing, can reduce excessive runoff. Groundwater is a viable option for Kenya, given its higher infiltration rate.
- **Tanzania:** Moderate runoff and infiltration indicate a balanced hydrological system. Investments in irrigation infrastructure could help maximize agricultural productivity. Watershed management and afforestation programs can further improve water retention.

4.5 Conclusions

The prevailing conditions between the investigated East and West African countries are different. East African countries show lower temperatures and higher precipitation. From the application of the method, we conclude that all case studies display water surplus. That means that on an intra-annual scale, soil moisture reaches its maximum holding capacity at the end of the year. A significant amount of water is stored in the aquifers, and a significant part of it becomes runoff. The case studies in East African countries (Kenya and Tanzania) revealed a greater potential for the exploitation of water resources. In the case studies of these two countries, the precipitation heights and infiltration percentages are higher, while evapotranspiration rates are relatively low compared to those of Ghana and Burkina Faso. This means that more significant amounts of water resources could be used to cover demands for anthropogenic activities, mainly for irrigation purposes. The method's output can inform water resources management decisions, agricultural planning, and climate change impact studies. It is particularly useful in the examined case studies, where water resources are critical and understanding seasonal and inter-annual variability is crucial for effective management.

Acknowledgments This research was funded by the EU H2020 EWA-BELT project [862848] "Linking East and West African farming systems experience into a BELT of sustainable intensification".

References

Acheampong, P. K. (1988). Water balance analysis for Ghana. *Geography, 73*(2), 125–131.

Alley, W. M. (1984). On the treatment of evapotranspiration, soil moisture accounting, and aquifer recharge in monthly water balance models. *Water Resources Research, 20*, 1137–1149.

Burak, S., Bilge, A. H., & Ülker, D. (2022). Thornthwaite's method for the computation of the water balance. In H. Gökçekuş & Y. Kassem (Eds.), *Climate change, natural resources and sustainable environmental management. NRSEM 2021. Environmental earth sciences*. Springer.

Castelino, J. B., & Khamala, C. P. M. (1979). The role of water resources in development. In *Proceedings of the 13th annual symposium of the east African academy*. September 1977.

Donker, N. H. W. (1987). WTRBLN: A computer program to calculate water balance. *Computational Geosciences, 13*, 95–122.

Fish, R. E. (2011). *Using water balance models to approximate the effects of climate change on spring catchment discharge: Mt. Hanang, Tanzania*. Michigan Technological University.

Hendrayana, H., Widyastuti, M., Riyanto, I. A., Nuha, A., Widasmara, M. Y., Ismayuni, N., & Rachmi, I. N. (2021). Thornthwaite and Mather water balance method in Indonesian tropical area. *IOP Conference Series: Earth and Environmental Science, 851*, 012011.

Karpouzos, D. K., Baltas, E. A., Kavalieratou, S., & Babajimopoulos, C. (2011). A hydrological investigation using a lumped water balance model: The Aison River basin case (Greece). *Water and Environment Journal, 25*, 297–307.

Koukouli, P. G., Georgiou, P. E., & Karpouzos, D. K. (2022). Assessing the hydrological effect of climate change on water balance of a River Basin in northern Greece. In *Research anthology on environmental and societal impacts of climate change* (pp. 817–839). IGI Global.

Liebe, J. R., Van De Giesen, N., Andreini, M., Walter, M. T., & Steenhuis, T. S. (2009). Determining watershed response in data poor environments with remotely sensed small reservoirs as runoff gauges. *Water Resources Research, 7*, 45.

Mammoliti, E., Fronzi, D., Mancini, A., Valigi, D., & Tazioli, A. (2021). WaterbalANce, a WebApp for Thornthwaite–Mather water balance computation: Comparison of applications in two European watersheds. *Hydrology, 2021*(8), 34.

McCabe, G. J., & Markstrom, S. L. (2007). *A monthly water-balance model driven by a graphical user interface*. US Geological Survey Reston.

Müller, M. D. S., Dourado-Neto, D., Timm, L. C., Reichardt, K., Sartori, F. F., & Felisberto, G. (2018). Climate analysis for agricultural improvement of the economic community of West African States according to Kppen and Thornthwaite. *African Journal of Agricultural Research, 13*(23), 1198–1212.

Napoli, A., Crespi, A., Ragone, F., Maugeri, M., & Pasquero, C. (2019). Variability of orographic enhancement of precipitation in the alpine region. *Scientific Reports, 9*(1), 13352.

Steenhuis, T. S., & Van Der Molen, W. H. (1986). The Thornthwaite – Mather procedure as a simple engineering method to predict recharge. *Journal of Hydrology, 84*, 221–229.

Thornthwaite, C. W., & Mather, J. R. (1955). *The water balance* (Vol. 8, pp. 1–104). Laboratory in Climatology, Johns Hopkins University.

Thornthwaite, C. W., & Mather, J. R. (1957). *Instructions and tables for computing potential evapotranspiration and the water balance* (Vol. 10, pp. 181–311). Laboratory in Climatology, Johns Hopkins University.

Internet Sources

https://aquastat.fao.org/climate-information-tool/
https://data.apps.fao.org/
https://earthwise.bgs.ac.uk/

Chapter 5
Heightening Sustainability Transition Through Business Transformation in the Fresh Produce Industry: Conceptual Framework and a Case Study

Samir Mili and Tasnim Loukil

5.1 Introduction

Developing more sustainable business models (BMs) has become crucial for the survival of companies and industries as environmental and social pressures (depleting natural resources, rising social inequalities) are increasing (Bocken et al., 2013). As a result, numerous companies have adapted their BMs to include environmental, social, and economic considerations (Broccardo et al., 2023), while business innovation for sustainability has received increased attention from researchers and practitioners (Yang et al., 2017). These demands for sustainability can be seen from a positive side as they push agri-food industries and firms to creatively incorporate efficient and effective innovations that help in preserving the environment and improving the social and economic resources. The engagement towards sustainability-oriented innovation will only be fruitful in the case where the changes are integrated and integral, which will affect the company and its larger stakeholder environment.

This strategy of sustainability-oriented innovation is focused on the inclusion of the triple bottom line (TBL) of corporate sustainability, encompassing the economic, the environmental and the social aspects. The integrated changes following this strategy will significantly lead to changes in organizations' BMs. Moreover, the volatility in consumer demand, the focus on healthy lifestyles—at least in Europe (Ngugi et al., 2020; Savelli et al., 2020; Tan et al., 2022; Rieger et al., 2023)—and the expanded attention paid to natural resources and animal welfare, as well as to workers' safety, represent additional motivations to shifting to sustainable BMs.

S. Mili (✉) · T. Loukil
Institute of Economics, Geography and Demography (IEGD), Spanish National Research Council (CSIC), Madrid, Spain
e-mail: samir.mili@csic.es

K. Mattas et al. (eds.), *Biodiversity in Agri-Food Systems*, Cooperative Management, https://doi.org/10.1007/978-3-032-06385-4_5

It should also be noted that in the past, businesses have been looking at sustainability issues from a narrow view where they focused more on the economic capital as the core of the organization and failed to reduce their impact on natural and social capitals. This has contributed to create economically sustainable companies in the short and medium term but a failure in the long term due to the non-substitutability of the three capitals (economic, social, environmental). Accordingly, failure to reduce the negative impact because of the non-integration of more ecological and social approaches consequently leads to a failure in the corporate sustainability.

The Spanish fruit and vegetable (F&V) industry is no exception to these ongoing patterns (Duque-Acevedo et al., 2022; Pozancos & Jiménez, 2019). Indeed, sustainability is at the heart of the challenges faced by the Spanish F&V industry (European Commission, 2023; Pozancos & Jiménez, 2019). Consumer demand for F&V is also increasingly influenced by sustainability issues (Camanzi et al., 2024; Gorynska-Goldmann et al., 2023). F&V products are one of the most crucial components of a healthy diet as they are abundant in essential nutrients, and play a vital role in preventing chronic diseases (European Commission, 2024; Wallace et al., 2020).

As for its socio-economic importance, the F&V sector in Spain is the most relevant agricultural sector in terms of production, added value, export propension, job creation, and continuance of the population in rural areas (Ministry of Agriculture, Fisheries and Food, 2023; De Castro, 2018). Spanish F&V production was 18.6 million tons in 2022 with an average production value of 20,450 million euros, which places Spain as one of the leading producers globally (Mercasa, 2023). F&V production represented 56% of the Spanish plant production in 2022, and 32% of the total agricultural production by value. In addition, 50% of the national production is destined for foreign markets, generating a total export value of 15,792 million euros in 2022 (Mercasa, 2023). From the organizational viewpoint, Fig. 5.1 summarizes the structure of the Spanish F&V supply chain including the main stages, participants, areas of activity and their interrelations.

From the policy perspective, the Spanish F&V sector should comply with the new, ongoing policy orientations resolutely more focused on sustainability, especially the new Common Agricultural Policy (CAP) 2023–2027 and its new instruments, like eco-schemes, the European Union (EU) Farm to Fork Strategy within the framework of the European Green Deal, the climate neutrality objective by 2050, the EU 2030 Biodiversity Strategy for protecting nature and ecosystems, and the EU fiscal policies to mitigate climate change. This is in addition to national and regional policies which in almost all cases are also more sustainability oriented. Moreover, as highlighted by Kenny et al. (2024), policy implementations should use more collaborative channels, and enhance the interdisciplinary connectedness of relevant actors within the fresh produce supply chain.

In this context, it is recognized that sustainability transition coupled with the adoption of supply chain innovations prompt new BMs in the F&V industry that enable the move from the standard economic-centered BMs towards more

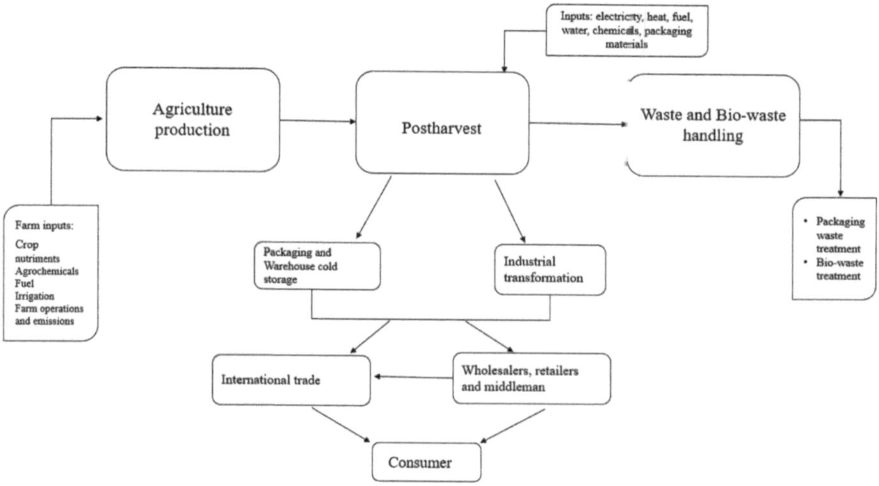

Fig. 5.1 Fruit and vegetable supply chain structure. (Own elaboration)

sustainable BMs explicitly integrating environmental and social concerns. However, sustainable business modelling is still poorly understood despite its importance to managers, scientists and policymakers confronting the complexities of business environment changes. In addition, while the sustainable business model (BM) is an emerging topic in the whole agri-food industry, still few tools exist to support companies and organizations implementing them (Geissdoerfer et al., 2016). There is a clear lack of knowledge about how sustainability objectives are virtually implemented in BMs of the F&V sector to deliver value on the triple bottom line of economic profit, environmental protection and social fairness.

Therefore, this study aims at contributing to literature in this topical area through exploring new approaches for the transition toward more sustainable BMs in the F&V industry. Based on the generic Triple Layered Business Model Canvas (TLBMC) proposed by Joyce and Paquin (2016), we develop a specific conceptual framework and perform a case study in the Spanish F&V industry, where the classical BM is innovated and transformed into a sustainable BM through the integration of sustainability goals. Data were gathered from different public and private sources to elucidate how sustainability concerns are integrated in BMs for F&V, with a focus on both the whole sector and a specific company as respective functional units of analysis. Ultimately, our study intends to test the pertinence and applicability of this model in the Spanish F&V industry, in order to portray the main business areas in the sector/company and identify those suited for potential innovations and improvements in terms of sustainable value creation and delivery. Applying the TBL analysis at the whole sector is also a previous step toward developing more tailored BMs at the company level.

5.2 Research Approach

The general approach followed in this study is based on the BM concept understood as a representation of the underlying logic of how a sector/company/organization creates and delivers value. One of the most renowned definitions of BM can be found in Teece (2010): "*a business model articulates the logic, data, and other evidence that support a value proposition for the customer, and a viable revenue and cost structure for the company that delivers that value*". The BM has also been used as a strategic positioning tool (e.g. Zott & Amit, 2007; Casadesus-Masanell & Ricart, 2010; Ferrer et al., 2022) which connects BMs to decisions in the supply chain. The design and implementation of the BM accounts for the activities that the company performs and the connections between these activities, leading to a BM structured in two parts: the actual actions that the company carries out (decision-making process) and the consequences of these actions (performance) (Ferrer et al., 2022).

The historical development of the BM concept has been related to a wide range of disciplines which produced a pluralism of perspectives, and a heterogeneous interpretation of the concept. Consequently, various endeavors were made in the literature to build up a synopsis of the different definitions and approaches already existing in research fields. Most extant studies are basically assigned to three theoretical approaches of the BM concept: technology-oriented, organization-oriented, and strategy-oriented approaches (Fig. 5.2).

Over the two past decades, the concept of sustainable BM has emerged and has been largely discussed in academic and business circles (Ulvenblad et al., 2019).

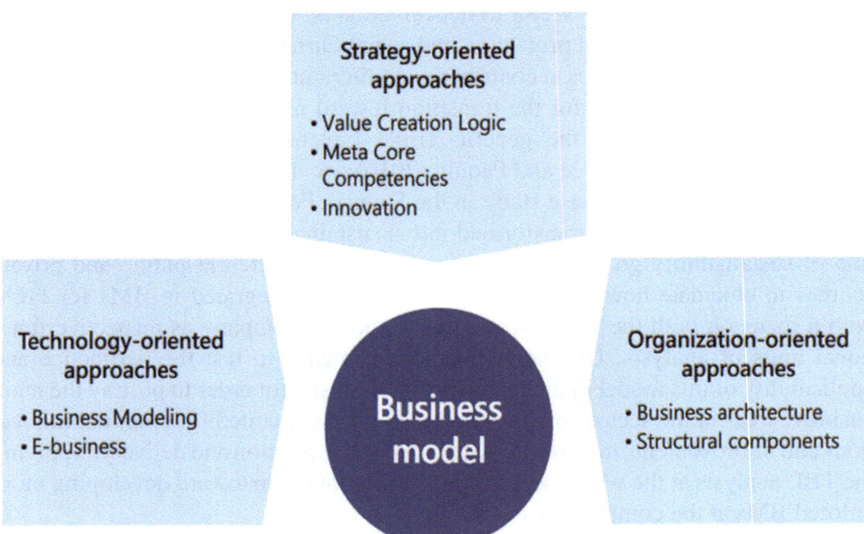

Fig. 5.2 Business models: three basic approaches (Wirtz, 2019)

One of the most cited definitions of sustainable BM is found in Bocken et al. (2013), wherein the sustainable BM represents "*Innovations that create significant positive and/or significantly reduced negative impacts for the environment and/or society, through changes in the way the organization and its value-network create, deliver and capture value…*".

In sustainable BMs, the standard and economic-centered BM is innovated and transformed into a sustainable BM through the integration of sustainability practices and concerns. Sustainable BMs are designed to integrate sustainability into companies (Broccardo & Zicari, 2020) while becoming a source of competitive advantage (Porter & Kramer, 2011). Researchers proposed them as alternatives to the traditional BMs where the major focus is on maximizing growth and revenue and minimizing cost, in order to consider the environmental and social dimensions of sustainability. Moreover, achieving a sustainable BM involves changes in business practices (Amit & Zott, 2012), and the involvement of different stakeholders (Sommer, 2012).

One of these alternatives is the TLBMC. This model is a novel framework that consistently adds two more layers to the classical economic BM canvas: the environmental layer based on a life cycle perspective, and the social layer based on a stakeholder perspective (Joyce & Paquin, 2016). Each layer has nine standard building blocks. Hence, the TLBMC is a framework following the dimensions of economic, environmental, and social performance representing the three pillars of sustainability (Malinauskaite & Jouhara, 2024; Purvis et al., 2019).

Admittedly, when applied in practice, the model should be conceptually adapted to the specific context of each study case and then empirically validated with the relevant data. This model also enables a structured representation of business activities and issues. For instance, our suggested TLBMC model facilitates a systemic representation of the very numerous and varied issues in the F&V supply chain (Tort et al., 2022), in such a way that they can be more understandable, visible and useful for business planning.

Other features of the TLBMC are: the use of inside-out approaches, which implies starting with the existing BM and then exploring the potential changes and improvements; the use of life cycle analysis (LCA) in the environmental layer to measure the environmental impacts through various indicators like CO_2 emissions, eco-systems quality, human health, resource depletion, water use; and a stakeholder approach in the social layer to evaluate social impacts using methods like social-LCA and Social Return on Investment, and implementation of ISO 26000 standards for social responsibility (identify key stakeholders and map them—through interviews, questionnaires, focus groups, showing the boundaries and indicators of economic, social and environmental impact analysis). The perspective of the whole approach is to balance the interests of all sector stakeholders rather than only maximizing gains for one participant. Economic impacts can be assessed by investment appraisal methods like Net Present Value, life cycle cost analysis, and profit margin (Ribeiro et al., 2018).

Regarding the empirical study carried out, the process has been straightforward. First, a literature review on both the F&V sector and BMs was performed. Then, data were collected from diverse public and private sources. The next step was the BM conceptualization through performing a conceptual framework based on the adaptation of the general TLBMC model to the specific context of the F&V sector. Subsequently, the applicability of the proposed conceptual framework has been tested in a real-world case study.

5.3 Results and Discussion

5.3.1 Building a Conceptual Framework Linking Business Models and Sustainability Factors in the F&V Industry

Based on the BM literature and sector review, we developed a conceptual framework establishing the conceptual link between sustainability issues and the different TLBMC canvases in the F&V sector. It should be pointed out that the step to building this tailored conceptual framework was the most difficult stage owing to the lack of previous research. The suggested conceptual framework was built through an adaptation of the generic TLBMC model proposed by Joyce and Paquin (2016) to the Spanish F&V sector and its specificities. It represents a meaningful tool for F&V companies seeking to innovate their BMs and to keep up to date with market trends, food security and consumer satisfaction through sustainability practices. It is composed of three different models that entail economic, environmental, and social perspectives of the fresh produce business. Each of the models is made up of nine components. The canvas models representing the three perspectives allow the firm to have a clear, overall and more dynamic vision of its business through the horizontal and vertical relations between the three layers and the nine blocks of each one with the purpose of increasing effectiveness and efficiency.

Table 5.1 illustrates the economic conceptual model in the F&V sector. The economic perspective is represented by the Business Model Canvas (BMC), where determining the key resources and the partners allows good identification of the value proposition needed to be generated for the target customers, the resources required to perform the key activities and those enabling cost savings. The BMC of the economic layer contributes to identifying the created value for customers and determining whether it satisfies their needs. From a pure economic perspective, the BMC is a tool that helps firms and industries to reduce their costs. Thus, the optimization of the proposed value can influence the cost structure. The first step to take is to identify all costs and expenditures associated with the business; with the realistic understanding of the cost, a clear and good BM can be formulated.

The next step is to list all the fixed and variable costs on the canvas which will help to clearly visualize if they follow a strategy of economy of scale (unit costs decrease as production increases) or economy of scope (costs are decreased by investing in businesses related to the core product). The revenue stream represents the income generated from each of the customer segments. Thus, generating income from revenue streams requires efficient and creative methods to attract different

Table 5.1 F&V economic conceptual model. (Own elaboration)

<table>
<tr>
<td rowspan="2">Key partners
Key partnerships for actors of F&V value chain towards sustainability (Ministry of Agriculture, other public agencies, input and service suppliers, research centers, etc.)</td>
<td>Key activities
Indispensable activities for implementing firms' value proposition considering the role of sustainable practices in value creation (sustainable production, logistics, marketing, packaging)</td>
<td rowspan="2">Value proposition
Structure of F&V value chain required to create and distribute offerings, and complementary assets needed to support position in the chain</td>
<td>Customer relationships
Engagement of consumers, type of relationships established and maintained by firms with customers (e.g. quality)</td>
<td rowspan="2">Customer segments
Market segments in national and international markets, priority segments (e.g. families, people with healthy lifestyle, people interested in environmental issues)</td>
</tr>
<tr>
<td>Key resources
Material and human resources necessary to create value for customers (labor, materials, land, capital, knowledge)</td>
<td>Channels
Appropriate distribution channels for F&V productions in national and international markets (retailers, wholesalers, direct sale, e-commerce, etc.)</td>
</tr>
<tr>
<td colspan="2">Cost streams
Cost structure of producing sustainable products (processing, marketing, materials to produce sustainable products, etc.)</td>
<td colspan="3">Revenue streams
Mechanisms by which F&V firms will be paid for the transition to sustainable models and for offering sustainably produced products (selling fresh F&V, organic and other sustainably produced F&V products, etc.)</td>
</tr>
</table>

customer segments and convince them to buy the company products and services. The profit calculated from the difference between revenue and cost, and the decision on how much the firm can invest and where to intervene, represents the final step to operate in a sustainable business.

In practice, the economic layer provides a structured representation of the economic aspects of the F&V sector business system, and the potential interconnections and impacts on value creation. As can be deduced from Table 5.1, some potential innovations and improvements could be explored, for instance in the areas of the choice of distribution channels, market segmentation, or enhancing consumer awareness of sustainable attributes to increase sales. In any event, for each potential innovation, costs and benefits should be evaluated systematically. Moreover, cost control in different areas of production, processing and marketing would potentially improve the financial situation.

The second model in the conceptual framework is the environmental BMC (Table 5.2), which is oriented further to the environmental costs and benefits generated by the F&V sector, and the reduced impact caused by the sector in this domain. It enables firms to better visualize the impact of the F&V supply chain on the environment, and to better mapping, showing where they need to reduce their impacts and where to focus when conducting environmentally oriented innovations. In this layer, the materials (main resources) represent the bio-physical stocks to produce the functional value through production processes. Another source to produce the functional value comes from the supplies and outsourcing; this source is considered essential but not core to the firm. After being processed, the product or the service is distributed by various means which are included in the distribution block, which also includes distances travelled and the weight of the shipped F&V. The shipped product to the final consumer needs other products and materials in order to be consumed, as well as maintenance and repair which are included in the use phase. The last phase of a product is the end-of-life where the consumer decides to end the consumption of the functional value. This phase supports the firm in discovering new methods to manage and reduce its environmental impact. The environmental

Table 5.2 F&V environmental conceptual model (own elaboration)

<table>
<tr>
<td rowspan="2">Supplies and out-sourcing
Materials and activities necessary to production processes other than “Materials” and “Production”</td>
<td>Production
Research and development (R&D) of innovative varieties. Variety license and growing.
Processing with latest technology generation.
Packaging.
Logistics</td>
<td rowspan="2">Functional value
Principal output of a product (e.g. sustainable fresh F&V, transformed products from F&V)</td>
<td>End of life
Last phase of the functional value after being used by the consumer/ end-user</td>
<td rowspan="2">Use phase
Essential resources employed by consumers when using F&V products (water for washing the products, energy, etc.)</td>
</tr>
<tr>
<td>Materials
Resources needed to provide the functional value considering the one with the highest impact (e.g. fertilizers)</td>
<td>Distribution
Trucks transport
Transoceanic boats
Logistic centers</td>
</tr>
<tr>
<td colspan="2">Environmental impacts
Ecological costs and negative externalities of activities of F&V companies (e.g. CO_2e, soil degradation)</td>
<td colspan="3">Environmental benefits
Positive ecological value created by F&V business activities and positive externalities through reduction of the environmental impact</td>
</tr>
</table>

BMC is also a tool used to locate the highest ecological cost caused by the different components of the F&V supply chain, and the environmental benefits generated.

Therefore, the environmental layer includes the risks and opportunities in terms of environmental value. For the F&V industry, it seems clear that several areas are suitable for potential innovations and improvements, for example, in the domains of carbon emissions, soil degradation, recycling, energy and water consumption, waste management, and pesticide and fertilizer management.

The third layer is the social BMC (Table 5.3). This layer is an extension of the original BMC with the adoption of the stakeholder approach. It clarifies the mutual effects between stakeholders and the company/organization, and it captures the impacts that derive from those relationships. The social layer is a visual tool to understand the virtual impact of the company/organization and where to innovate its actions and BM toward the improvement of its social value creation potential. The social value creation is a key part of sustainability-oriented organization's mission

Table 5.3 F&V social conceptual model (own elaboration)

Local communities Stakeholders and communities with whom F&V companies and organizations have social relationships. Each community should be considered as a different stakeholder with its specific needs and realities	**Governance** Organizational structures and decision-making policies. Specifies who are the stakeholders to engage with and the way to proceed (e.g. contractual arrangements, transparency in decision-making, etc.)	**Social value** Creating benefits for firm's stakeholders and society (improving health and quality of life of F&V consumers, etc.)	**Societal culture** Culture of responsibility (e.g. fight against corruption and fraud, improve trust and transparency) Employment Culture of respecting persons and their rights	**End-users** Final persons who buy/use/consume the value proposition. Special attention to needs of end-users and contribution to their quality of life (e.g. product taste, nutritional benefits, health benefits)
	Employees Role of employees as core organizational stakeholders, as well as their working conditions		**Scale of outreach** Depth and breadth of the developed relationships between F&V organizations and their stakeholders through actions over time at different scales (local, regional, national, international)	
Social impacts Social costs of F&V companies and organizations			**Social benefits** Positive social value created from sustainable practices of F&V firms and organizations	

where focus is on creating social benefits for their local communities, employees, and stakeholders in general. Based on their governance structure, policies and decision-making processes, organizations define which stakeholders to engage with and the way to proceed.

In addition, not only governance but also the societal culture is being taken into consideration when social value is created, as well as the positive benefits for the end-users and the different stakeholders. Thus, an organization is creating value and waiting for positive feedback from the end-users who acquire the value proposition. Put differently, an organization seeks to contribute to improve the quality of life of the end-user and not only to satisfy his/her needs. In this perspective, an organization builds various relationships with different stakeholders through the actions it takes over time and this is called the scale of outreach. The latter describes the relationships of an organization in terms of the geographical outreach and the long-term maintenance of these relations, as well as the manner of integrating a societal culture and differences when creating social value. As in the economic and environmental layers, the social layer presents the social cost of an organization on its different stakeholders e.g. employees, local suppliers.

In short, the social layer portrays the different variables, factors and interrelations for each one of the nine building blocks. In the F&V industry, major potential social benefits can be identified, for instance in the areas of working conditions, stakeholder commitment, and nutritional and health benefits. Hence, innovating and improving these areas can provide highly appreciated social value.

5.3.2 *Case Study*

Based on the suggested conceptual framework depicted above, we performed a case study assessing how a Spanish F&V company integrates sustainability in its BM and the resulting outcome on value creation performance.

Company Presentation We selected the Spanish AMC group which is divided into two main holdings: AM Fresh and AMC Natural Drinks (AMC Group, 2022). The motivation for choosing this firm was related to its sustainable performance in the Spanish F&V sector. Indeed, besides the pure economic aspects, this business holding moved its focus to the environmental and social aspects. Moreover, the fact that this firm operates internationally also improved the data availability.

The AMC group is a third-generation family firm that occupies the third position in terms of international Spanish food sales. It is a leader in developing and licensing vegetable varieties, production and marketing of fruits, flowers, ice-cream and natural drinks with high added value. The goal of being an international benchmark in the food industry is shown through its different activities and approaches that represent its mission and commitment to sustainability, innovation, transparency, and society. Through the sustainable process, the company also seeks to offer consumers an added value towards healthier lifestyles.

AMC Natural Drinks is made up of more than 20 companies and is dedicated to the production of chilled juices, smoothies, and vegetable milks. With a turnover of 628 million euros in 2022, the company stands out for its positioning in the international market, supplying its products to more than 70 supermarket chains, mainly in Europe. This company is vertically integrated, as it processes its raw materials in more than 20 countries, being one of the largest F&V juicers in Europe. This corporation packages all types of drinks in its five international plants. The company is commercially present on the five continents, and has its own commercial offices in the main European countries.

AM Fresh is an international benchmark in the agricultural market, with a presence and leadership in the vegetable varieties sector. AM Fresh is present in a large number of countries with more than 20 operational centers and 2500 ha of production located on three continents and has recently carried out acquisitions that has allowed strong growth in America, Europe and Asia.

SWOT and TOWS Analysis A SWOT analysis has been conducted in order to size up the company's resource capabilities and deficiencies, as well as the opportunities and threats. The results of the analysis are schematically illustrated in Fig. 5.3. The analysis illustrates the continuous improvements of the firm in terms of strategic management and sustainable ways of production. The company with its different affiliations is going along with the technological developments which it applied in favor of sustainable production for both fresh F&V and other processed products.

With respect to the internal factors (strengths and weaknesses), the AMC group already has a strong sustainability-oriented BM innovation; their engagement

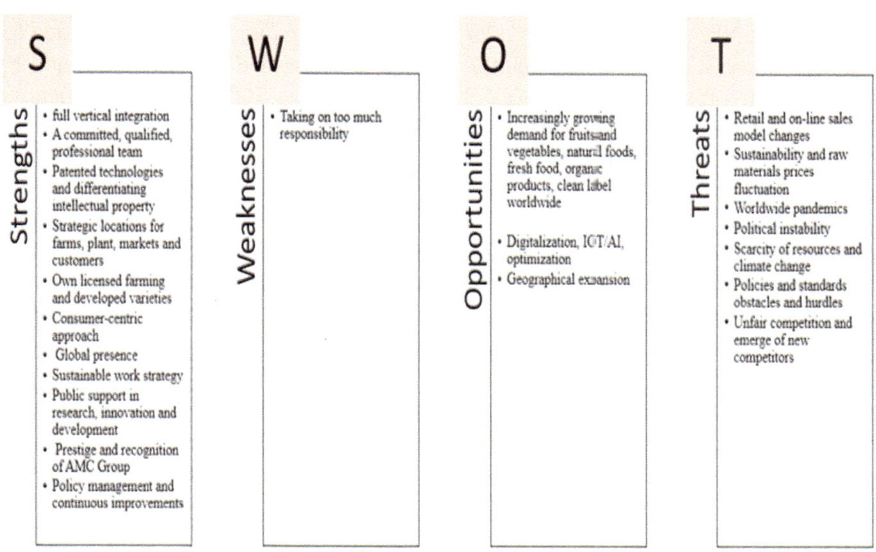

Fig. 5.3 SWOT analysis of the AMC group. (Own elaboration)

towards the three aspects of sustainability is indicated with the different innovations used in production and the attention paid to details like policy management, branding and the constant development strategy. As for the external factors (opportunities and threats), the company is taking advantage of the outside (external) opportunities, which is illustrated in their strengths; having satisfied consumers with all the various products they offer, the business shows that it is taking advantage of the expansion of demand for sustainable food. Awareness of all the threats outside the business and all the changes that are taking place, from the sanitary and geopolitical risks to the emerging competitors, is one of the main points occupying the company's close attention.

The AMC group and its different affiliations has moved toward being a sustainable business. The group is in constant evolution and ongoing development, managing to balance internal and external factors of success. The employment of the well-defined strategies is paying off for the AMC group and its two main holdings, where the commitment to work and the good understanding of F&V markets on national and international levels are determining factors of success. The company is benefiting from the opportunities of the growing market for sustainable and organic F&V products. It continuously operates based on its strengths to increase sales, to retain new customer segments and to develop internationally. Among the internal factors that create strength are the constant research and development and the attention paid to production systems, supply chain management, customer needs and policies. Following a full vertical integration permits the group to attentively manage each step of the supply chain from the development of the varieties to consumer satisfaction through innovative technologies in production along with sustainable practices, i.e. a platform for customers as part of the consumer-oriented strategy, with new technologies for industrial transformation of F&V for various products. Furthermore, the AMC Natural Drinks worked on being internationally certified for quality, where it has been constantly audited and certified by customers and by international organizations such as Sure Global Fair (SGF), BRC Certification Body, and the European Fruit Juice Control System.

The TOWS analysis of the AMC group provides the connection of the different components from the SWOT analysis to better visualize and capture value. The SO represents the max-max situation where the company uses the strengths to take advantage of the opportunities. In this perspective, the group can use its prestigious and recognized brand to introduce new products in the market. The company can exploit its research centers to innovate either in the process of production or in the product itself through the optimization of energy use; for example, it takes advantage of its research centers to manage to reduce energy consumption and waste through the IoT and the digitalization of different parts of the production process.

Another strength is the consumer-centric approach. With the opportunity of the geographical expansion and the increasing demand on organic food, the AMC group can customize each type of product to the various types of consumers worldwide. Moreover, with the licensed farming the group possesses, the research centers where new varieties are constantly developed, and the strategic location of the plant and farms, the company can develop varieties with specifications adapted to different customers from different parts of the world.

Meanwhile, the ST allows exploiting the strengths to avoid threats. It is another part of the TOWS analysis based on the components of the SWOT analysis. One of the main threats to the company is the changes in policies and the imposed policies from national, European and international parties. To deal with these threats, the company works on policy management lines to adapt the firm's activities to the existing regulations.

The scarcity of resources and climate change is a major threat warning food-producing firms including the AMC group. To deal with this threat, the group can deploy the different research centers it has to innovate new ways of production and varieties that do not harm nature. It can also apply technological patents to reduce resource exploitation and waste and increase production systems that reduce pollution and the impact of climate change. Employing sustainable work strategies, sustainability-oriented BM, and a market global presence, it can reduce the impact of climate change and scarcity of resources.

The changes in retail sales models is a significant challenge for some firms where sales are increasing online. In this area, the AMC group has a dynamic BM and a sustainable strategy favoring online sales with the goal of reducing air pollution and GHG emissions from logistics.

The WO represents overcoming weaknesses by using the opportunities. In this way, the strategy of applying external factors to control the internal factors and to solve the weaknesses of the group is a great option, where the AMC group can use technological development and optimization, as well as global expansion, to reduce the responsibility taken to provide healthier and more sustainable products.

The WT reflects the challenge for the AMC company not only to overcome the weaknesses but also to minimize the threats on its business. In this sense, getting over the resource scarcity and climate change is a real challenge facing the whole F&V industry.

It should be noted that SWOT and TOWS analyses are highly recommended for strategic planning. As illustrated, SWOT analysis identifies the key internal and external factors that are responsible for the company's current position. The clear and understandable presentation of the AMC group's strengths, weaknesses, opportunities, and threats makes it easy to communicate the opportunities to be seized and the different levels of the supply chain to be ameliorated (Sarsby, 2016). Using SWOT and TOWS also helps the company to expand its product categories, the sustainable practices it is using, and the flexible BMs used with retailers, among other objectives.

Applicability of the TLBMC Perspective The application of the suggested conceptual model to the studied company produced a series of components portraying the three sustainability dimensions, allowing the firm to have a clear, comprehensive and more dynamic vision of its business with regard to sustainability objectives.

Overall, the economic layer represents the traditional BMC focusing on the financial issues of the group and how to generate profit while reducing costs. The social layer investigates the relationships between the AMC group and its stakeholders from suppliers to consumers. This layer aims to analyze the major social benefits and impacts from using sustainable practices. The environmental layer is based on a holistic view of the products of the company from a life cycle perspective, and it is built with the objective of creating more environmental benefits from introducing sustainable practices.

Putting sustainability at the forefront of the AMC group's strategic business plans and implementing corporate sustainability have been offering numerous advantages to the group and its affiliations. This implementation creates a positive environmental return supporting the prevention of climate change and bolsters the reduction of GHG emissions and organic and non-organic waste, responsible use of non-renewable energies, and biodiversity conservation. These sustainable practices also present social and economic benefits for the company.

Developing new varieties as a sustainable practice to prevent climate change increases productivity and reduces costs. Reducing the energy and resource consumption leads to further efficiency and cost savings. Moreover, reducing the waste and being transparent enhance the credibility and help to build strong relationships with the different stakeholders of the company. The sustainable practices lead to building more trust and brand reputation, and they also increase the competitive advantage of the company. The strong reputation of the company through sustainable practices and the certification obtained from national and international organizations can lead to an increase in sales and improvement in the position of the company compared to other firms.

However, implementing more sustainable practices can also be a motive to increase production costs, resulting in higher prices. This increase affects the purchasing power of consumers and their willingness to pay for the AMC group products.

The commitment to sustainability also drives the company to follow a customer satisfaction strategy. This engagement leads in turn to more brand recognition and an increase in sales, considering that a satisfied consumer plays the role of communication channel where he/she can promote the product, by giving positive feedback. Furthermore, the sustainable practices improve consumer health, through the development of new varieties and innovative healthy products, as well as the reduction of pollution and better waste management (Râpă et al., 2024).

Interestingly, the application of the triple bottom line involving the AMC group case study has led to constructive conclusions about the effectiveness and efficiency of the introduced sustainable practices. The results of the AMC group have shown that the TLBMC represents a visual tool that provides a holistic view of the business from three different perspectives: the economic, the environmental and the social. Thus, it facilitates the uncovering and better understanding of the company BMs and the sustainability-oriented innovation.

Similarly, the applied framework provides greater clarity and a better understanding of the different elements of the company's business and the created value under the TBL perspective, but also the negative impacts in the three sustainability areas. In addition, it highlights the connections leading to more collaborative approaches between the functional and departmental sections of the company. As a visual tool, the TLBMC raises an awareness and captures the changes resulting from introducing sustainable practices in the BMs, and allows users to explore the modifications and the impact of these practices. Moreover, the TLBMC qualifies users to balance between impacts and benefits across the three axes of sustainability. For instance, in the case of the F&V industry, the introduction of sustainable practices such as the carbon footprint reduction generates an increase in the economic cost, but at the same time it creates social and environmental benefits (reduction of GHG emissions, climate change mitigation, human health, among others), which can be rewarded by consumers and increase revenues.

5.4 Conclusions and Implications

The present study attempted to explore the TBL approach and its applicability in the F&V industry in the context of the transition from traditional to sustainable BMs. It also examined the sustainable practices implemented in the F&V sector and how they can contribute to this transition. This study contributes to existing literature mainly by increasing a conceptually and empirically validated understanding about how innovating in BMs impacts the F&V industry. Owing to the scarcity of previous research on the topic, the present study is essentially explorative in nature.

The introduction of this innovative approach to the F&V sector will explicitly raise awareness of the importance of sustainable practices. Additionally, it is a tool that will help to comply with the policies and regulations of the EU that aim to reduce the negative impact on the environment, and to meet the actual human needs without compromising the ability of future generations to meet their own needs.

The case study has shown that the use of a flexible pattern-matching approach can be useful not only for researchers but also for managers. Research through a real-life case study can also be more convincing and better understood by practitioners. This technique presents a high degree of flexibility as it starts with a theoretical pattern to end up with three layers adapted to the company.

The analysis carried out suggests that the three layers of the model should be considered simultaneously, i.e. a holistic view should be present when renovating and designing BMs. The interactive relationship and coherence between the three layers should be highlighted in order to achieve better, more useful analysis. Indeed, the case study performed demonstrates that the key to success relies on the equal importance given to the three layers, involving relevant and coherent innovations in the economic, environmental, and social dimensions.

Findings indicate that the transition from traditional to sustainable BMs is a complex but fundamental step in the sustainability path. The suggested conceptual framework emphasizes the process of innovating in BMs and the role of sustainable practices adopted by F&V firms leading to sustainable transition. It helps to get the whole picture of the impacts associated with production and marketing in the F&V

industry, which would be a prior step that supports companies in the sector to embed sustainability concerns in their BMs.

Results indicate that, overall, sustainability goals can be represented appropriately within the TLBMC framework. The model can be an effective, accessible tool to expand an economic-centered approach of standard BMs by developing environmental and social layers. It facilitates the identification of potential innovations and improvements while contributing to promote a business culture where innovation and sustainability are explicitly addressed when developing new BMs. In this model, sustainability innovations are much better communicated than in the standard model, which can significantly change the value proposition and differentiate the offerings (product, price and service), with consequent gains in terms of competitive advantage.

It is shown that sustainable practices in the fresh produce industry generate sustainable value with positive impacts, not only on the environment but also on the economic and social status of the sector's participants. Indeed, sustainable practices can lead to growth in sales and profits since many consumers are increasingly oriented towards sustainably produced products (e.g. organic and other eco-friendly products), while environmental benefits can create further value for F&V businesses, stakeholders and society as a whole. Social benefits can also be generated in several areas such as employment, engagement with public policies and societal concerns, trust, and transparency. Sustainability actions also increase reliability and reputation which are key levers for value creation and competitive advantage.

It can be argued that a change of paradigm is needed, moving from a component to systems thinking, because sustainable value-creation processes depend greatly on vertical and horizontal coherence within and between layers involving many interrelated variables and factors. So far, little consideration has been given to this needed shift in perspective in BM research, and we think that this contribution is a step in this direction.

It should be noted that the suggested conceptual framework can presumably be applicable to different categories of enterprises, since it visually communicates the company's activities, provides a holistic view of the business in question, and details activities based on sustainability TBL. However, the model is very challenging to apply in a comprehensive way at the sector level, especially when there is a very wide range and a high complexity of products and activities. Hence, the TLBMC would be, a priori, more suited to smaller functional units (SMEs/subsidiaries/divisions) with a small number of products.

Data limitations also represent a serious challenge for model applicability, especially data for the environmental and social layers, since there is no established system for collection and analysis of this type of data. Another issue is the scarcity of information on internal policies of companies and their organizations. Furthermore, several intricacies have been encountered when analyzing the social layer as some components were hard to identify and measure, hence the need for more methodologies and tools for measuring social impact.

Future research should include collecting further primary data from major experts and stakeholders in the F&V sector using structured surveys and interviews. Longitudinal analysis would also be needed to better understand the evolution of different components of the model over time, and to identify potential patterns and trends in the data. Triangulation is also needed to enhance the validity, accuracy and consistency of the results. Another important issue is how to deal with external shocks, uncertainty and factor volatility inside and outside the F&V supply chain, which pushes to changing BMs without a monitoring system. In addition, more company case studies should be carried out in order to obtain more factual outcomes and solid conclusions.

References

AMC Group. (2022). AMC Group – Estado de información no Financiera Consolidado 2021/2022 (Consolidated non-financial information statement 2021/2022). https://amcnaturaldrinks.com/wp-content/uploads/2023/01/EINF_GROUP 2022_c.pdf

Amit, R., & Zott, C. (2012). Creating value through business model innovation. *Sloan Management Review, 53*(3), 41–49.

Bocken, N., Short, S., Rana, P., & Evans, S. (2013). A value mapping tool for sustainable business modelling. *Corporate Governance, 13*(5), 482–497.

Broccardo, L., & Zicari, A. (2020). Sustainability as a driver for value creation: A business model analysis of small and medium enterprises in the Italian wine sector. *Journal of Cleaner Production, 259*, 120852.

Broccardo, L., Zicari, A., Jabeen, F., & Bhatti, Z. A. (2023). How digitalization supports a sustainable business model: A literature review. *Technological Forecasting and Social Change, 187*, 122146.

Camanzi, L., Ahmadi Kaliji, S., Prosperi, P., Collewet, L., El Khechen, R., Michailidis, A., Charatsari, C., Lioutas, E. D., De Rosa, M., & Francescone, M. (2024). Value seeking, health-conscious or sustainability-concerned? Profiling fruit and vegetable consumers in Euro-Mediterranean countries. *British Food Journal, 126*(13), 303–331.

Casadesus-Masanell, R., & Ricart, J. (2010). From strategy to business models and onto tactics. *Long Range Plannning, 43*(2–3), 195–215.

De Castro, L. (2018). Spanish fresh fruits and vegetables sector. In *OECD fruit and vegetables scheme. Proceedings of the 18th meeting of the heads of National Inspection Services* (pp. 20–30) https://www.oecd.org/agriculture/fruitvegetables/publications/Proceedings-2018-HNIS-meeting.pdf

Duque-Acevedo, M., Belmonte-Ureña, L., Terán-Yépez, E., & Camacho-Ferre, F. (2022). Sustainability and circularity in fruit and vegetable production. Perceptions and practices of reduction and valorization of agricultural waste biomass in South-Eastern Spain. *Journal of Environmental Management, 316*, 115270.

European Commission. (2023). *EU agricultural outlook for markets, 2023–2035*. Available at: https://agriculture.ec.europa.eu/data-and-analysis/markets/outlook/medium-term_en

European Commission. (2024). *Food-based dietary guidelines recommendations for fruit and vegetables*. https://knowledge4policy.ec.europa.eu/health-promotion-knowledge-gateway/food-based-dietary-guidelines-europe-table-3_en

Ferrer, J. R., García-Cortijo, M. C., Pinilla, V., & Sebastián Castillo-Valero, J. (2022). The business model and sustainability in the Spanish wine sector. *Journal of Cleaner Production, 330*, 129810.

Geissdoerfer, M., Bocken, N., & Hultink, E. (2016). Design thinking to enhance the sustainable business modelling process – a workshop based on a value mapping process. *Journal of Cleaner Production, 135*, 1218–1232.

Gorynska-Goldmann, E., Murawska, A., & Balcerowska-Czerniak, G. (2023). Consumer profiles of sustainable fruit and vegetable consumption in the European Union. *Sustainability, 15*, 15512.

Joyce, A., & Paquin, R. (2016). The triple layered business model canvas: A tool to design more sustainable business models. *Journal of Cleaner Production, 135*, 1474–1486.

Kenny, D., Dermiki, M., & Britton, F. (2024). Implementing Ireland's food vision 2030 within the fresh produce sector: An investigation of the barriers and enablers using systems thinking principles. *Sustainability, 16*, 7237.

Malinauskaite, J., & Jouhara, H. (2024). Chapter 3: Sustainable business models: Components, drivers and barriers. In *Sustainable energy technology, business models, and policies: Theoretical peripheries and practical implications* (pp. 67–97). Elsevier.

Mercasa. (2023). Alimentación en España: Producción, industria, distribución y consumo (Food in Spain: Production, industry, distribution and consumption). https://www.mercasa.es/wp-content/uploads/2023/11/ALIMENTACION-EN-ESPANA _2023.pdf

Ministry of Agriculture, Fisheries and Food. (2023). *Información del sector de frutas y hortalizas en España* (Information on the fruit and vegetable sector in Spain). https://www.mapa.gob.es/es/agricultura/temas/producciones-agricolas/frutas-y-hortalizas/informacion_general.aspx

Ngugi, I. K., O'Sullivan, H., & Osman, H. (2020). *Consumer behaviour in food and healthy lifestyles: A global perspective*. CABI.

Porter, M. E., & Kramer, M. R. (2011, January–February). *Creating shared value*. Harvard Business Review.

Pozancos, J. M., & Jiménez, B. (2019). Los desafíos del sector español de frutas y hortalizas en un entorno cambiante (The challenges of the Spanish fruit and vegetable sector in a changing environment). *Distribución y Consumo, 158*, 5–12.

Purvis, B., Mao, Y., & Robinson, D. (2019). Three pillars of sustainability: In search of conceptual origins. *Sustainability Science, 14*(3), 681–695.

Râpă, M., Darie-Niță, R. N., & Coman, G. (2024). Valorization of fruit and vegetable waste into sustainable and value-added materials. *Waste, 2*, 258–278.

Ribeiro, I., Sobral, P., Peças, P., & Henriques, E. (2018). A sustainable business model to fight food waste. *Journal of Cleaner Production, 177*, 262–275.

Rieger, J., Freund, F., Offermann, F., Geibel, I., & Gocht, A. (2023). From fork to farm: Impacts of more sustainable diets in the EU-27 on the agricultural sector. *Journal of Agricultural Economics, 74*, 764–784.

Sarsby, A. (2016). *SWOT analysis*. Lulu.com.

Savelli, E., Francioni, B., & Curina, I. (2020). Healthy lifestyle and food waste behavior. *Journal of Consumer Marketing, 37*(2), 148–159.

Sommer, A. (2012). *Managing green business model transformations*. Springer.

Tan, B. C., Lau, T. C., Sarwar, A., & Khan, N. (2022). The effects of consumer consciousness, food safety concern and healthy lifestyle on attitudes toward eating "green". *British Food Journal, 124*(4), 1187–1203.

Teece, D. (2010). Business models, business strategy and innovation. *Long Range Planning, 43*(2–3), 172–194.

Tort, Ö. Ö., Vayvay, Ö., & Çobanoğlu, E. (2022). A systematic review of sustainable fresh fruit and vegetable supply chains. *Sustainability, 14*(3), 1573.

Ulvenblad, P. O., Ulvenblad, P., & Tell, J. (2019). An overview of sustainable business models for innovation in Swedish agri-food production. *Journal of Integrative Environmental Sciences, 16*, 1–22.

Wallace, T. C., Bailey, R. L., Blumberg, J. B., Burton-Freeman, B., Chen, C. O., Crowe-White, K. M., Drewnowski, A., Hooshmand, S., Johnson, E., Lewis, R., Murray, R., Shapses, S. A., & Wang, D. D. (2020). Fruits, vegetables, and health: A comprehensive narrative, umbrella review of the science and recommendations for enhanced public policy to improve intake. *Critical Reviews in Food Science and Nutrition, 60*(13), 2174–2211.

Wirtz, B. W. (2019). *Digital business models. Concepts, models, and the alphabet case study*. Springer.

Yang, M., Vladimirova, D., & Evans, S. (2017). Creating and capturing value through sustainability. *Research-Technology Management, 60*(3), 30–39.

Zott, C., & Amit, R. (2007). Business model design and the performance of entrepreneurial firms. *Organization Science, 18*(2), 181–199.

Chapter 6
The Status and Role of Smallholder Cow's Milk Production in the Supply Chain in Hungary

Judit Bernadett Vágány and Szilvia Erdeiné Késmárki-Gally

6.1 Introduction

Smallholder cow's milk production has always played an important role in Hungary. In the rural areas of the country, there was hardly a settlement that could not supply itself with fresh cow's milk. Cattle farming was based mainly on the dual-purpose Magyartarka breed. The country's favourable geographical location enabled families to produce the fodder needed to feed their animals, thus ensuring their livelihood. Generations succeeded each other in heading the farms. Farming has contributed greatly to the rural population's ability to survive. Later on, the sale of the milk that was not used in the production of raw cow's milk, was replaced, delivering it to the dairy parlour, from where it was transported in refrigerated tanks by the dairy processing plants in refrigerated trucks (Vágány, 2008; Szakály, 2001).

The situation in Hungary has changed significantly in the past decades. Small-scale milk production has been replaced by large-scale units. A significant concentration has taken place in the sector and this process is expected to continue in the future (OTP Agrár, 2021). Small-scale farms have also experienced difficulties in the process of generational change: farming is no longer seen as attractive for the younger generation, which wants more effortless work with fewer constraints

J. B. Vágány (✉)
Department of Economics and Business Studies, Faculty of Commerce, Hospitality and Tourism, Budapest Business University, Budapest, Hungary

Department of Management and Leadership, Budapest University of Economics and Business, Faculty of Management, Budapest, Hungary
e-mail: vagany.judit@uni-bge.hu

S. E. Késmárki-Gally
Faculty of Business, Communication and Tourism, Institute of Economics and Finance, Budapest Metropolitan University, Budapest, Hungary
e-mail: sgally@metropolitan.hu

K. Mattas et al. (eds.), *Biodiversity in Agri-Food Systems*, Cooperative Management, https://doi.org/10.1007/978-3-032-06385-4_6

(Farrell et al., 2022; Vágány, 2023). Dairy farms have disappeared, making it almost impossible to get the raw material to the dairy processor. The remaining smallholder farms focus mainly on meeting local needs. In this study we will examine the situation of smallholder cow's milk production.

6.2 Literature Review

In the literature review, we present the situation of raw cow's milk production in the largest dairy-producing countries of the European Union and in the neighbouring countries of Hungary, followed by a more detailed discussion of the Hungarian situation.

The international situation is based on studies in the Web of Science and Scopus databases, and Eurostat is used to provide statistical data. The Hungarian situation is presented using data from the most important Hungarian scientific journals, the Hungarian Central Statistical Office and the Institute of Agricultural Economics.

6.2.1 Raw Cow's Milk Production in the European Union

Raw milk production has always played an important role in the life of the European Union. According to the European Parliament (2018), it accounts for more than 12% of total agricultural production.

The EU is currently the second largest milk producer in the world, with a raw milk production of around 161 million tonnes (Eurostat, 2022).

In 1984, a milk quota system was introduced to regulate production, with the aim of controlling the quantity of cow's milk produced. The quota system was the only element that had a direct impact on the dairy farming sector until the introduction of the milk subsidy in 2004. The milk quota system basically guaranteed the establishment and maintenance of market balance by regulating the quantities of raw milk produced and sold, together with other market regulation instruments. The Union has made milk producers who produce milk in excess of the national quota pay the costs of placing or storing cow's milk on the market by imposing a levy (MVH Notice No 177/2011 (19.12.2011)).

The EU milk quota system was in place until 1 April 2015. The functioning of the quota system, its abolition and the period since then have been examined by several authors in their academic work (Bouamra-Mechemache et al., 2008; Kempen et al., 2011; Čechura et al., 2021; Groeneveld et al., 2016; Jongeneel & Gonzalez-Martinez, 2022).

According to Eurostat data, following the abolition of the quota system, the two largest milk-producing countries in the EU (Germany and France) did not see a significant increase, but other countries were able to take advantage of the abolition of the quantitative barrier of milk quotas and increase their output (e.g. Italy, the Netherlands, Spain and Belgium) (Fig. 6.1).

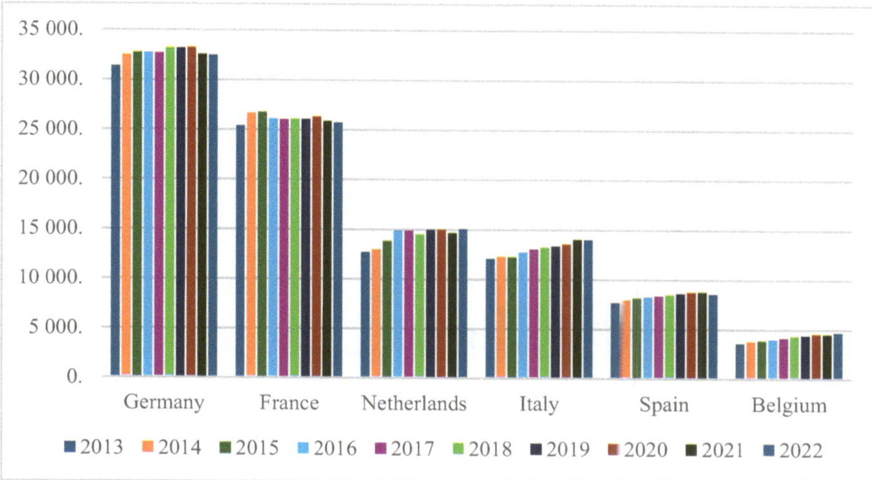

Fig. 6.1 Total raw milk available on farms in the main EU dairy-producing countries (1000 t). (Source: Own editing, based on Eurostat, 2024)

As our study focuses mainly on the Central and Eastern European region, we were also interested to see what the impact of quota abolition was in this region. Poland is indisputably the largest milk producing country in the region. It has been the winner of the abolition of the quantitative limit and has significantly increased its milk production. An increase was also recorded in Austria, the Czech Republic and, to a lesser extent, Hungary. However, the situation was less favourable in the other countries of the region, with a slight decrease in Romania and stagnation in Slovakia and Slovenia (Fig. 6.2).

Moving further down the supply chain, and taking a brief look at the processing industry, we see that of the raw milk produced (estimated by Eurostat at 160 million tonnes in 2022), a total of 9.8 million tonnes were used on dairy farms, with the rest going to dairy processors. For the purposes of our topic, it is important to examine the fate of the milk left on farms and how farms used it.

The data show that some of the milk remaining on farms was consumed by the producer and his/her family or sold directly to the consumer. Other possible uses include the use of milk as animal feed and direct processing (e.g. cheese, butter, etc.). Of the 149.9 million tonnes of milk delivered to dairies, 145.6 million tonnes were cow's milk, while the rest was sheep's milk (3 million tonnes), goat's milk (2.5 million tonnes) and buffalo milk (0.3 t) (Fig. 6.3).

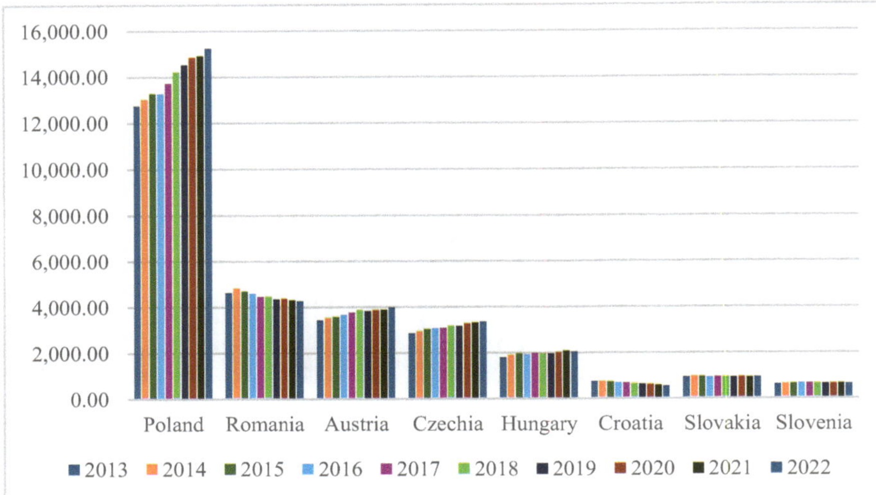

Fig. 6.2 Total raw milk available on farms in Central and Eastern European countries (1000 t). (Source: Own editing, based on Eurostat, 2024)

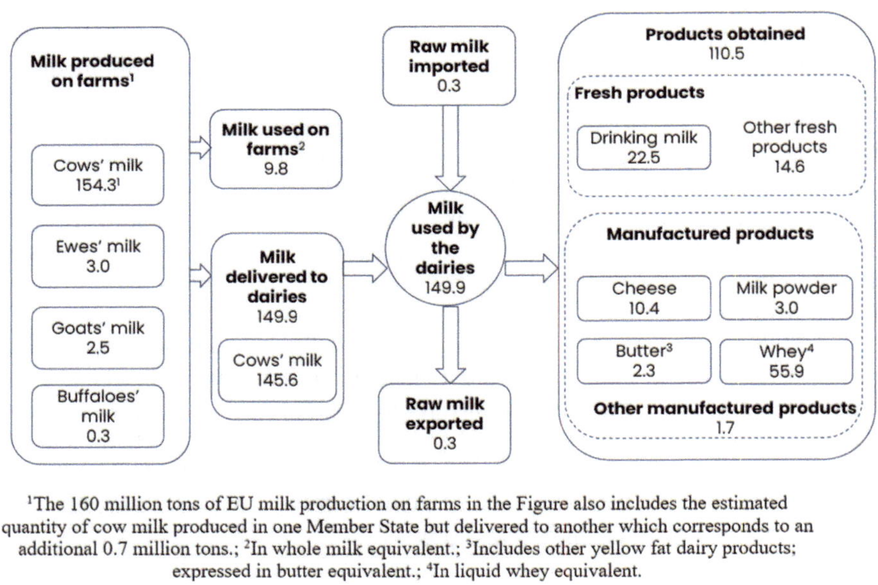

Fig. 6.3 Milk production and consumption in the European Union in 2022. [1]The 160 million tons of EU milk production on farms in the Figure also includes the estimated quantity of cow milk produced in one Member State but delivered to another which corresponds to an additional 0.7 million tons.; [2]In whole milk equivalent.; [3]Includes other yellow fat dairy products; expressed in butter equivalent.; [4]In liquid whey equivalent. (Source: Eurostat, 2024)

6.2.2 *Production of Raw Cow's Milk in Hungary*

In the 1980s and even in the early 1990s, in smaller villages, a significant proportion of families kept cows. The cattle herd reached the level of 2 million heads, of which nearly half were cows (Fig. 6.4).

Basically, dual-purpose Magyartarka stocks were predominant on farms, while Holstein-Friesian stocks were also becoming increasingly important (Buzás & Szabó, 2011; Vágány et al., 2003).

The numbers ranged from one or two cows to hundreds. These farms varied widely in terms of both technology and quality, but the quantity of milk produced was at an unprecedented level in Hungary in the 1980s (Fig. 6.5).

A well-established supply chain guaranteed the raw materials for dairy processors. Small producers collected cow's milk mainly in milk barns, from where it was transported in refrigerated tankers. There were also some who sold raw milk door-to-door or processed it and made cheese, butter or sour cream for the local population. These farms usually met local needs with their products (Vágány, 2008; Gál et al., 2013).

In the 1990s, however, large numbers of people (skilled and unskilled) who had worked in agriculture left the sector in Hungary, and on many farms—due to low wages—only the lower skilled remained, often out of necessity, to work with the animals.

Farmers across Europe are struggling with the problems of climate change (Fróna et al., 2021), with low profitability, changing farm structure and successful generational change (Farrell et al., 2022; Suess-Reyes & Fuetsch, 2016; Tomchuk et al., 2018; Shamsuddoha et al., 2023).

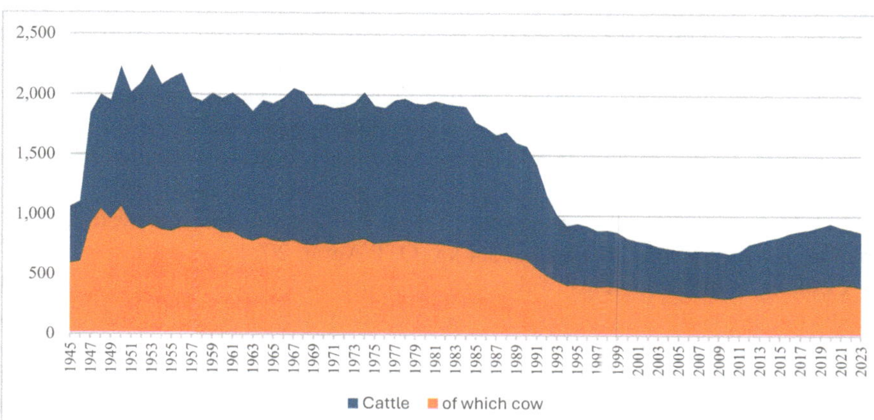

Fig. 6.4 Size of cattle and cow stock between 1945 and 2023 (thousands). (Source: Own editing, based on data from the Hungarian Central Statistical Office, 2024a)

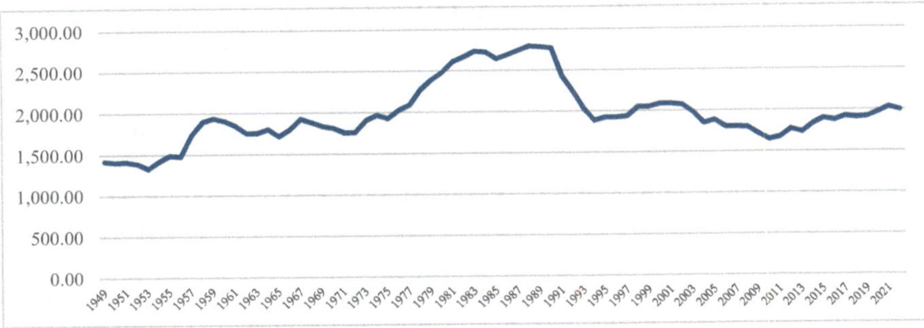

Fig. 6.5 Cow milk production in Hungary 1949–2022. (Source: Own editing, based on data from the Hungarian Central Statistical Office, 2024b)

The age of workers in the agricultural sector has steadily increased and the number of small farms has decreased (Kolosi & Sági, 1996). Nowadays, it is almost a curiosity to find small-scale cattle farming in a municipality.

The number of raw cow's milk-producing farms in Hungary is very heterogeneous by county (see Fig. 6.6). The counties of Hajdú-Bihar, Pest and Bács-Kiskun stand out, as these regions play a significant role in agricultural production. Their more favourable geographical conditions allow for a high level of agricultural activity.

According to the National Food Chain Safety Office, a total of 1274 farms produced milk in January 2024. The number of registered dairy farms is 80, of which 32 do not buy raw milk and process their own milk; 15 have only one supplier and only 33 farms have relevant purchases (National Food Chain Safety Office, 2024).

Dairy production in Hungary is concentrated regarding international correlation, with large farms accounting for the majority of domestic production. Between 2011 and 2021, the average number of cows per farm increased significantly from 354 to 442, which means that the concentration of dairy cows has been steadily increasing and we expect this trend to continue in the future. Farms with more than 500 heads account for almost 60% of the dairy cow population, but these larger farms represent only 2% of the total number of farms. Specific production indicators have also increased steadily: monthly milk production per cow has risen from 1293 to 1679 kilograms in 10 years, an increase of 30%. At the same time, the average milking rate increased from 26.1 kg to 33.4 kg (OTP Agrár, 2021).

Looking at the economics of dairy cow farming in Hungary, it can be seen that production costs have increased steadily over the period under study (Table 6.1). The same is true for the cost price. The unit income per litre was only 7.73 HUF in 2021, which can be considered very low. The drastic reduction in the number of small farms did not start in the period under review, but their number shrank even further during this specific time, partly due to negative economic effects. The profitability of the sector is so low that it is no longer an option for farms with only a few cows to continue operating.

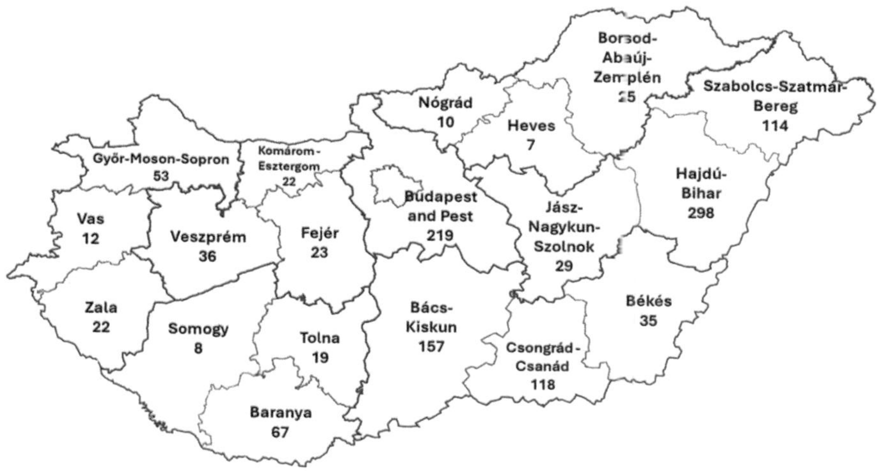

Fig. 6.6 Distribution of raw cow's milk-producing farms by county in Hungary in January 2024 (pieces). (Source: Own editing, based on data from the National Food Chain Safety Office, 2024)

Table 6.1 Economic parameters of dairy cow husbandry

Title	Dairy cow husbandry				
	2017	2018	2019	2020	2021
Milk production cost (HUF/quart)[a]	758,037	824,603	864,995	812,549	901,820
Average yield (litres per hectare)	7917	8530	8589	8372	7835
Cost price (HUF/litre)	89.90	91.23	95.69	92.5	111.23
Sales price (HUF/litre)	95.57	95.14	99.34	106.38	118.97
Specific income (HUF/litre)[b]	5.67	3.91	3.65	13.88	7.73
Sectoral result (HUF/quarter)[c]	264,294	240,841	267,127	334,793	252,123
Sectoral result per 100 HUF of production cost (HUF)	34.87	29.21	30.88	41.20	28.07

Source: Institute of Agricultural Economics (2020, 2021, 2022, 2023)
[a]Average number of persons per year
[b]Without subsidies
[c]Including subsidies, 1 EUR = 395 HUF

6.2.3 Requirements for the Chemical and Physical Properties of Raw Cow's Milk

The requirements for the chemical and physical properties of raw cow's milk are largely aimed at guaranteeing the true, original condition of the milk. The hygiene standards are designed to ensure that raw cow's milk is suitable for human consumption. It is also important to consider that there is a close link between hygiene properties and milk quality (Szakály, 2001).

Let's first look at the factors affecting agricultural production in general (Table 6.2). Several physical, biological, socio-economic, cultural, legal and customer conditions can be identified in the production of agricultural products. For a better overview, both the factors related to livestock production and those affecting crop production are considered, since the output of crop production can be an input (e.g. feed) for livestock production.

As can be seen from Table 6.2, there are a number of factors that play a significant role in the production of agricultural products that farmers cannot prevent or control or can only partially prevent or control.

In the case of crop production, the physical factors cannot be controlled or can only be controlled at high cost, while in the case of livestock production, it is the weather, temperature and breeding cycle that can be controlled in an artificial environment in intensive farming. In extensive farming, risks are beyond the control of the farmer.

A number of studies on agricultural production and food safety, including factors affecting human health and consumers' knowledge of food safety (Tomchuk et al., 2018; Al Mamun et al., 2023; Röhr et al., 2005; Kasza et al., 2024a; Ssubi et al., 2024; Duan et al., 2024; Bozsik et al., 2023; Abdallah et al., 2023; Kocsis & Kuslits, 2019; Galloa et al., 2020). Several hazards complicate the production of raw cow's milk in regards to quality control and food safety.

Table 6.2 Factors affecting agricultural production

	Crop production	Livestock
Physical	Solar radiation Temperature Precipitation Soil Topography	Weather Temperature
Biological	Pests and pathogens Cultivated plants (species, variety) Growing technology Natural vegetation	Health condition Pathogens Support technology Estrous cycle Species/variety
Socio-economic	Population The social structure Economic environment Technical conditions Commercial background Workforce	Population The social structure Economic environment Technical conditions Commercial background Workforce
Cultural	Expertise Ideology, religion Historical traditions	Expertise Ideology, religion Historical traditions
Legal/customer	Plant health standards Contractual agreements Food law requirement Standard specifications	Animal health requirements Contractual agreements Food law requirement Standard specifications

Source: Own editing, based on Altieri (1989)

Table 6.3 General and plant-specific hazard identification

Microbiology	Biological	Chemistry	Physical
Salmonella ssp., E. coli, S. aureus, Mycobacterium tubercolosis/bovis, Brucella, Listeria monocytogenes, Pseudomonas aeruginosa/ fluorescens, Campylobacter jejuni, Shigella sonnei, Bacillus cereus, Faecal streptococci	Pathogenic and facultatively pathogenic pathogens, Somatic cells, Insect/bug/rodent droppings	Aflatoxin from feed, Pesticide residues from feed, Hormone residue, Antibiotic—residue from medication, Oil, grease—from machinery technology, Cleaning, disinfectant residue—technology on machines, Nitrite, nitrate, heavy metals from water	Animal hair, Human hair, Dust, litter, excrement, Pieces of metal, rubber, wood, tiles from the environment Other foreign material from equipment

Source: Own compilation

We have identified the most common hazards to be avoided and minimised by farmers. Regarding hazards, microbiological, biological, chemical and physical hazards are discussed in more detail (Table 6.3).

There are several options available to farmers to minimise the risks. The first is to follow conscious management principles, for which the application of good practices (GMP, GHP, etc.) and HACCP principles provide an excellent framework. It is also vital that farmers have the latest information on food safety.

The income-generating capacity of agriculture is low. This is particularly true for the production of raw cow's milk (see Table 6.1). A food safety system could most likely be set up and operated by a farm only with the help of a consultant, which, depending on the size of the farm, the number of employees and the complexity of the activity, could have significant costs. Smaller farms will most likely not be able to afford the additional costs of setting up and running a system. (The situation is better for larger farms. The HACCP, ISO 9001, etc., which are currently in operation, have been introduced mainly on these).

Risk communication also has an impact on the reputation of food safety authorities (Kasza et al., 2024b), so effective communication between all actors in the food chain is essential.

6.3 Methods

We used a mixed methodology in our study. In our secondary research, we used international and Hungarian literature sources to present the situation of the dairy sector in Hungary and its neighbouring countries, which are the most important from the market point of view. In the framework of our primary research, we conducted semi-structured interviews with the managers of small dairy farms in a small region and presented a potential HACCP process through the example of a medium-sized farm in order to gain insight into the process and critical control points of raw cow's milk production by applying the principles of HACCP.

The interviewees were selected from farmers in a specific small area in the Pest County. An important criterion for the selection of farmers was to interview farms already included in our previous research (2006). The interviews were conducted between May and June, 2024.

The number of respondents was planned to be larger, but during the contact process, it became clear that many of the farms we wanted to include no longer existed, so the final number was reduced to 7 people who still run working farms. A further 14 were selected from among farmers who had recently stopped farming.

Our questions focused on two major issues for working farms: (1) an assessment of the current situation and the issue of generational change; (2) an assessment of the attitude towards HACCP and food safety. The managers of farms no longer in operation were asked about the reasons for the closure. As for the medium-sized farm, there were 120 dairy cows (and their offspring) at the time the flow chart was prepared. Here, we focused on the milk production process and aimed to show a good practice.

6.4 Results

6.4.1 Results of Interviews with Managers of Smallholder Cow's Milk Farms

Over the past 15 years, the number and size of raw cow's milk farms in Hungary has completely changed. A strong concentration has started, with more and more small farms abandoning cow's milk production. The profitability of the sector is so low that it is no longer an option at all for farms with a few cows to maintain activity. We were aware of this when we started our research. However, the facts we learned in preparing our primary research came as a surprise. In one municipality, for example, out of the 21 small farms which were included in the research that formed the basis of the study (carried out in 2006), only 1 farm is still producing raw cow's milk. The situation is not much better in other municipalities. We also contacted 14 farmers who were still producing cow's milk during the original survey period but have since ceased their activities.

The following answers were given to our questions on the reasons for the farm closures:

- the farmers have grown old and can no longer take care of the farm,
- the family member who cared for the animals has died and the surviving family member is no longer able to care for the animals alone,

- the younger generation no longer wants to deal with animals,
- the dairy has closed, so farmers no longer have anywhere to sell their milk
- production costs (feed, veterinary medicines, etc.) have increased enormously, and they receive so little money for their milk that it is no longer worth keeping the animals.

They thought a lot about how to continue their activities more economically. The possibility of the farmers joining forces to set up a milk processing plant, where they could produce higher added value was raised; the milk produced could be processed into bagged milk and sold to the inhabitants of the sub-region. Unfortunately, this initiative was also abandoned due to mistrust, lack of cohesion, insufficient support for the small producers' cooperative, and the additional investment required from its members to implement it.

Farmers produce raw cow's milk mainly to meet local demand. Only 2 out of the 7 farms surveyed are also involved in processing and produce cheese, yoghurt, and cottage cheese. They sell their products on small farmers' markets.

The farms keep Holstein-Friesian cows, and in the summer, they keep the animals extensively on pasture near their farm. The pastures are owned (3) or rented (4) by the farms. The average age of the farmers surveyed is high at 65 years, but it is promising that 4 out of the 7 farms surveyed are planning to change generations or have already started the process.

We then asked our interviewees about food safety. Their answers show that they are aware of food safety rules. They follow the news on dairy production and are well-informed in this respect.

Because they sell their products directly, they cannot afford to have any problems with the cow's milk produced, as one farmer put it. However, no HACCP system is in place. We also asked our interviewees about the reasons for this. Their answers showed that the farmers interviewed were aware of the HACCP method. In their opinion, they would see an improvement in the following areas if it was introduced:

- The organization would maintain documented procedures that provide for appropriate actions to identify and correct the cause of detected nonconformities, prevent their recurrence, and return the process or system to a controlled state after a nonconformity is detected.
- The farm would be able to take measures to manage non-compliant products to prevent them from entering the food chain unless it can be ensured that the food safety hazards concerned have been reduced to a specified, acceptable level.
- The user of the HACCP system can provide credible evidence of compliance during routine or ad hoc official controls.
- The introduction of the system could improve the "order" in the production of the product.
- It can improve the quality of the product produced.
- The amount of waste can be reduced.
- Costs can be reduced by rationalising production.
- However, producing good quality milk does not depend on having a certificate, but on applying a quality approach.

The arguments against the establishment and use of HACCP, together with common shortcomings and problems, are listed below:

- difficulties in separating clean and contaminated zones due to the design of the plant,
- rodent control with the assistance of an unofficial exterminator,
- the management of hazardous waste,
- the verifiability of the calibration of measuring instruments,
- the attitude of small farm managers towards the implementation and operation of HACCP (unnecessary, useless extra work and documentation).

Our research shows that in one- or two-person or family-run businesses, HACCP is essentially a formal system, with the same people responsible for overseeing and monitoring the system and for carrying out the operational tasks.

6.4.2 Example of a Medium-Sized Farm

The medium-sized economy examined in our study is not common in Hungary. In terms of equipment and innovative approach, it is larger than other farms of similar size.

The technology of dairy cows is modern, and some technological elements are new. Production is managed by a computerised farm management system which, in addition to recording production and reproductive biology data, also controls the computerised feeding of the cows. Since 1999, an activity measurement system has been in operation, which allows a better assessment of the reproductive biological status of the cows and a more reliable determination of the time of insemination. In 2002, a digital weighing system was installed in the corridor from the milking parlour to the barn and several improvements have been noticed since then (Székely, 2004).

The dairy farm has two buildings housing around 100 Holstein-Friesian dairy cows and their female offspring (heifer calves, heifer heifers). The most important production building on the farm is the cowshed. At the front of the building there is the 2×5 herringbone milking parlour, equipped with state-of-the-art milking equipment (individual identification, milk volume and flow rate measurement, recording of individual treatments, automatic milking cup removal). The 2500 litre milk cooler is sufficient for the hygienic storage of milk produced during the twice daily milking. An automatic cleaning system ensures that the milk lines are kept clean. A signal from the milking parlour is used to select the animals to be treated by means of an automatic gate, which can also be used to measure body weight using a digital scale. In the veterinary treatment room, it is possible to carry out hoof care, inseminations and minor veterinary procedures.

The cows are housed in a free-range system in cubicles. The bulk feed is distributed by a feeding trolley on the road in front of the covered feeding area along the two long walls of the building, while the bulk feed is distributed by automatic feeders located in the building. The slurry is removed using a wing-blade system and a water-saving process. The slurry storage tank holds 6 months' worth of slurry, which is spread by a slurry spreader tanker onto an appropriately selected field. The animals are free to choose their location between the pens, the feeding area or the paddock system, except in winter. The young cattle and dry cows are moved to pasture.

The herd bred here is free from brucellosis, leucosis and tuberculosis. Their condition is carefully monitored periodically and, in the manner, as required by law; any positive individuals are isolated from the herd. In order to maintain the healthy status, the current number of animals is supplemented only by the own breeding stock and no animals from other breeding stock may be brought onto the holding. The herd also participates in the exemption programmes for other infectious diseases in cattle breeding.

The farm has its own land; the aim is to supply the cattle with self-produced (low-cost) bulk feed. There is no mixing of abomasums; the necessary ready-mixed abomasums are supplied from mixing plants. Breeding is done by artificial insemination. The farm managers strive to implement breeding methods that continuously improve the genetic value of the herd and provide the opportunity to improve the quantity and quality parameters of the final product (raw cow's milk).

The farm decided to introduce HACCP back in September 2001. Their decision was motivated by several reasons:

- They were extremely open to any new initiative.
- They wanted their already high-quality milk to continue to be in the extra quality category; they saw the introduction of HACCP as a way to further ensure this.
- It was hoped that the introduction and certification of HACCP would give the farm a competitive advantage over other dairy farms that had not implemented a quality assurance system.
- By providing a steady supply of extra quality milk, they hoped to achieve a more favourable bargaining position, which would translate into better prices and better perception.
- It was expected that in the future, all commodity-producing farms would be obliged to use the method.
- They wanted to extend the existing management system on the farm with quality assurance. This is how the HACCP method became the HACCP system.

The HACCP system was introduced in January 2002 and subsequently audited in April 2002.

The process of milk production is shown in the following figure (Fig. 6.7), with the critical points marked.

Generally speaking, even before the introduction of HACCP, the farm paid great attention to the production of good quality milk. They also made the wages of cow workers dependent on the quality of the milk produced, making them more interested in paying attention. The high-quality milk was delivered directly to a dairy processor by road. Thanks to conscious and quality-oriented farming, the production of high-quality milk on the farm is guaranteed.

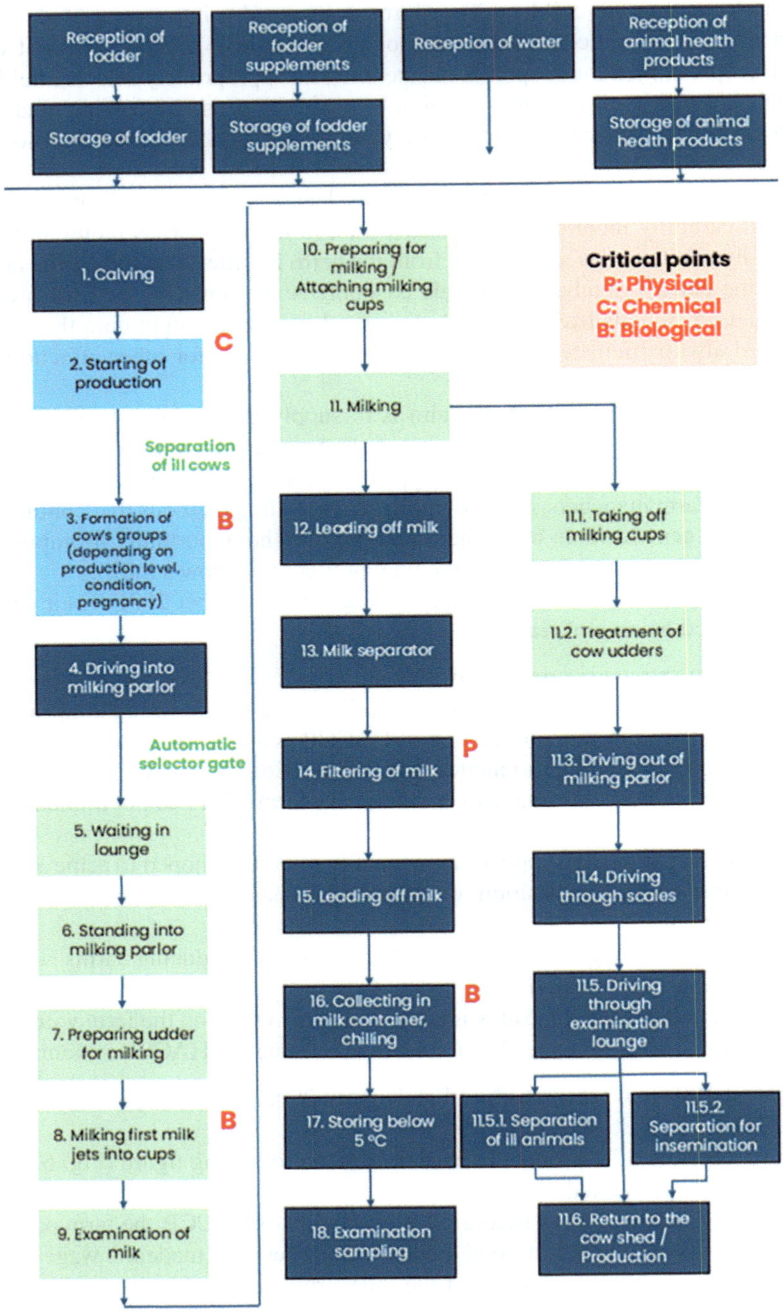

Fig. 6.7 Critical regulatory points for milk production in the medium-sized farm under study. (Source: Own editing, based on Vágány (2008) and Lencsés et al. (2014))

6.5 Conclusions

Small-scale cow's milk farms in Hungary produce high quality milk that meets local needs. In addition to the sale of milk, artisanal cheese, butter and yoghurt are also predominantly produced and sold. Their knowledge of food safety is good, they comply with the rules and, since they sell their products directly, they cannot afford to cause any health problems with their products, as this would risk their survival.

But the sector's profitability is so low that it is no longer an option for farms with a few cows to remain in business unless they can keep local customers willing to pay more than the average price of raw milk for good quality milk.

Their access to the supply chain has deteriorated considerably. One of the reasons for this is that, according to NÉBIH data from January 2024, of the 80 registered milk processing plants, 32 do not buy raw milk and they process their own milk; 15 have only one supplier, and only 33 have relevant purchases.

The current situation is therefore not favourable for small-scale cow milk farms. However, their survival is important, in particular to meet local needs. The average age of farmers is high, but the chances of generational change on farms are low. According to farmers, the younger generation does not want to be involved in livestock and dairy farming, as they prefer easier, less demanding work, which may predict a further decline in the number of small farms.

Overall smallholder cow's milk production is expected to shrink further in Hungary. The remaining small farms are expected to meet the local demand for raw cow's milk and artisanal products, as is the case today.

References

Abdallah, M. B., Farkas, M., Vasa, L., & Lakner, Z. (2023). Food security of Tunisia: Comprehensive analysis of a composite index. *Acta Polytechnica Hungarica, 20*, 10 pp. 199–216. https://doi.org/10.12700/APH.20.10.2023.10.12

Al Mamun, A., Prasetya, T. A. E., Dewi, I. R., & Ahmad, M. (2023). Microplastics in human food chains: Food becoming a threat to health safety. *Science of the Total Environment, 858*(Part 1), 159834. https://doi.org/10.1016/j.scitotenv.2022.159834

Altieri, M. A. (1989). *Agroecology: The science of sustainable agriculture*. Colorado. Westview Press a Division of Harper Collins Publishers.

Bouamra-Mechemache, Z., Jongeneel, R., & Réquillart, V. (2008). Impact of a gradual increase in milk quotas on the EU dairy sector. *European Review of Agricultural Economics, 35*(4), 461–491. https://doi.org/10.1093/erae/jbn044

Bozsik, N., Cubillos, T. J. P., Stalbek, B., Vasa, L., & Magda, R. (2023). Food security management in developing countries: Influence of economic factors on their food availability and access. *PLoS One, 17*: 7 paper: e0271696, 24. 2023. https://doi.org/10.1371/journal.pone.0271696.

Buzás, G., & Szabó, F. (2011). Economic evaluation of milk production of cattle breeds. *Farming, 55*(2), 2011. http://www.gazdalkodas.hu/index.php?l=hu&p=cikk&cikk_id=848

Čechura, L., Žáková Kroupová, Z., & Benešová, I. (2021). Productivity and efficiency in European milk production: Can we observe the effects of abolishing milk quotas? *Agricultures, 11*(9), 835. https://doi.org/10.3390/agriculture11090835

Duan, K., Onyeaka, H., Pang, G., & Meng, Z. (2024, December). Pioneering food safety: Blockchain's integration in supply chain surveillance. *Journal of Agriculture and Food Research, 18*(2024). https://doi.org/10.1016/j.jafr.2024.101281

European Parliament. (2018). The EU dairy sector. *Briefing*. https://www.europarl.europa.eu/RegData/etudes/BRIE/2018/630345/EPRS_BRI(2018)630345_EN.pdf

Eurostat. (2022). *Milk and milk product statistics*. https://ec.europa.eu/eurostat/statisticsexplained/index.php?title=Milk_and_milk_product_statistics

Eurostat. (2024). *How much milk does the EU produce?* https://ec.europa.eu/eurostat/web/products-eurostat-news/w/edn-20240531-1

Farrell, M., Murtagh, A., Weir, L., Conway, S. F., McDonagh, J., & Mahon, M. (2022). Irish organics, innovation and farm collaboration: A pathway to farm viability and generational renewal. *Sustainability, 14*(2022), 93. https://doi.org/10.3390/su14010093

Fróna, D., Szenderák, J., & Harangi-Rákos, M. (2021). Economic effects of climate change on global agricultural production. *Nature Conservation, 44*, 117–139. https://doi.org/10.3897/natureconservation.44.64296

Gál, T., Nagy, L., David, L., Vasa, L., & Balogh, P. (2013). Technology planning system as a decision support tool for dairy farms in Hungary. *Acta Polytechnica Hungarica, 10*, 8 pp. 231–244. https://doi.org/10.12700/APH.10.08.2013.8.15

Galloa, M., Ferrarab, L., Calogeroc, A., Montesanod, D., & Naviglio, D. (2020). Relationships between food and diseases: What to know to ensure food safety. *Food Research International, 137*(2020). https://doi.org/10.1016/j.foodres.2020.109414

Groeneveld, A., Peerlings, J., Bakker, M., & Heijman, W. (2016). The effect of milk quota abolishment on farm intensity: Shifts and stability. *NJAS – Wageningen Journal of Life Sciences, 77*, 25–37. https://doi.org/10.1016/j.njas.2016.03.003

Hungarian Central Statistical Office. (2024a). *Cattle, pigs, horses, sheep, buffaloes, donkeys, mules and goats*. https://www.ksh.hu/stadat_files/mez/hu/mez0027.html

Hungarian Central Statistical Office. (2024b). *Production of milk, eggs, wool, feathers and honey*. https://www.ksh.hu/stadat_files/mez/hu/mez0034.html

Institute of Agricultural Economics. (2020). *Agricultural Statistics Pocketbook, 2019*. https://www.aki.gov.hu/termek/hungarian-food-and-agricultural-statistics-2019/

Institute of Agricultural Economics. (2021). *Agricultural Statistics Pocketbook, 2020*. https://www.aki.gov.hu/termek/hungarian-food-and-agricultural-statistics-2020/

Institute of Agricultural Economics. (2022). *Agricultural Statistics Pocketbook, 2021*. https://www.aki.gov.hu/termek/agricultural-statistics-pocketbook-2021/

Institute of Agricultural Economics. (2023). *Agricultural Statistics Pocketbook, 2022*. https://www.aki.gov.hu/termek/agricultural-statistics-pocketbook-2022/

Jongeneel, R., & Gonzalez-Martinez, A. R. (2022). The role of market drivers in explaining the EU milk supply after the milk quota abolition. *Economic Analysis and Policy, 73*(2022), 194–209. https://doi.org/10.1016/j.eap.2021.11.020

Kasza, G., Izsó, T., Kunszabó, A., Szakos, D., Lénárt, I., Bozánné Békefi, E., Vongxay, K., Popp, J., Oláh, J., Lakner, Z., Nugraha, W. S., Bognár, L., & Süth, M. (2024a). Consumer knowledge, attitudes, and behaviour regarding food safety – A perspective from Laos. *Heliyon, 10*(9), e29725. https://doi.org/10.1016/j.heliyon.2024.e29725

Kasza, G., Izso, T., Langsrud, S., Vrbos, D., Veflen, N., Ueland, Ø., Scholdererf, J., Münter, L., Csenki, E., Szakos, D., Ramsay, J., & Süth, M. (2024b). Institutional food safety risk communication – A self-evaluation tool and its interpretation. *Trends in Food Science & Technology, 150*, 104594. https://doi.org/10.1016/j.tifs.2024.104594

Kempen, M., Witzke, P., Domínguez, I. P., Jansson, T., & Sckokai, P. (2011). Economic and environmental impacts of milk quota reform in Europe. *Journal of Policy Modeling, 33*(1), 29–52. https://doi.org/10.1016/j.jpolmod.2010.10.007

Kocsis, T., & Kuslits, B. (2019). Multidimensional labelling: Closing the sustainability information gap between producers, consumers and sustainability science in the food sector. *Periodica Polytechnica Social and Management Sciences – 2019, 27*(1), 9–16. https://doi.org/10.3311/ppso.12594

Kolosi, T., & Sági, M. (1996). Systemic change and social structure. In R. Andorka, T. Kolosi, & G. Vukovich (Eds.), *Social report 1996* (pp. 149–197). TARKI.

Lencsés, E., Kovács, A., Dunay, A., & Mészáros, K. (2014). *Changes to the HACCP system in a dairy farm due to the installment of an automatic milking system.* Stowarzyszenie Ekonomistów Rolnictwa I Agrobiznesu 313 Roczniki Naukowe. 2014. XVI. 6.

MVH. (2011, December 19). The agricultural and rural development agency, Hungary. Notice No 177/2011

National Food Chain Safety Office, Hungary. (2024). *Approved Dairy Farms, 2024.* https://www.nebih.gov.hu/documents/10182/1075873486/Tejtermgazd__2024_09_03.pdf/68d629b5-0a54-c287-3171-1b95b7bd5baf?t=1725367092443

OTP Agrár. (2021). *Tejtermelés és tejipar – nem bizakodhatunk csak a támogatásban.* https://www.otpagrar.hu/Velemeny/Reszletek?id=105296

Röhr, A., Lüddecke, K., Drusch, S., Müller, M. J., & Alvensleben, R. V. (2005). Food quality and safety – Consumer perception and public health concern. *Food Control, 16*(2005), 649–655. https://doi.org/10.1016/j.foodcont.2004.06.001

Shamsuddoha, M., Nasir, T., & Hossain, N. U. I. (2023). A sustainable supply chain framework for dairy farming operations: A system dynamics approach. *Sustainability, 2023*(15), 8417. https://doi.org/10.3390/su15108417

Ssubi, J. A., Mukisa, I. M., & Muyanja, C. K. (2024). Knowledge, attitudes and practices of fresh Nile perch value chain handlers towards food safety requirements in Uganda. *Heliyon, 10*(10), e31432. https://doi.org/10.1016/j.heliyon.2024.e31432

Suess-Reyes, J., & Fuetsch, E. (2016). The future of family farming: A literature review on innovative, sustainable and succession-oriented strategies. *Journal of Rural Studies, 47*(2016), 117–140. https://doi.org/10.1016/j.jrurstud.2016.07.008

Szakály, S. (2001). *Tejgazdaságtan*. Dinasztia Kiadó.

Székely, C. S. (2004). *Integrated family economic model – Józsefmajor*. Agroinform Publishing House.

Tomchuk, O., Lepetan, I., Zdyrko, N., & Vasa, L. (2018). Environmental activities of agricultural enterprises: Accounting and analytical support. *Economic Annals-XXI, 169*(1–2), 77–83. https://doi.org/10.21003/ea.V169-15

Vágány, J. (2008). *Economic impact and development potential of quality cow's milk production in Hungary* (PhD thesis). SZIE Gödöllő, 2008. https://archive2020.szie.hu/file/tti/archivum/vagany_j_dissz.pdf

Vágány, J. B. (2023). Generational change in family farms and the situation of arable land, as a factor of production, in Hungary. *Acta Polytechnica Hungarica, 20*, 10 pp, 127–143. https://doi.org/10.12700/APH.20.10.2023.10.8. (1785-8860 1785-8860)

Vágány, J., Dunay, A., Székely, C. S., & Pető, I. (2003, November 28–29). *Development and introduction of HACCP system in Józsefmajor experimental and demonstration farm, a dairy farm for fresh milk production*, IAMO Workshop, Halle., ISBN 3-9309270-0-8.

MIX
Papier aus verantwortungsvollen Quellen
Paper from responsible sources
FSC® C105338

If you have any concerns about our products,
you can contact us on
ProductSafety@springernature.com

In case Publisher is established outside the EU,
the EU authorized representative is:
Springer Nature Customer Service Center GmbH
Europaplatz 3, 69115 Heidelberg, Germany

Printed by Libri Plureos GmbH
in Hamburg, Germany